ENVIRONMENTAL SCIENCE, ENGINEERING AND TECHNOLOGY

WEATHER SATELLITES

DEVELOPMENT PROGRESS AND CONTINGENCY GAP ISSUES

ENVIRONMENTAL SCIENCE, ENGINEERING AND TECHNOLOGY

Additional books in this series can be found on Nova's website under the Series tab.

Additional e-books in this series can be found on Nova's website under the e-book tab.

ENVIRONMENTAL SCIENCE, ENGINEERING AND TECHNOLOGY

WEATHER SATELLITES

DEVELOPMENT PROGRESS AND CONTINGENCY GAP ISSUES

SHERRIL G. RYDER
EDITOR

nova publishers
New York

For permission to use material from this book please contact us:
Telephone 631-231-7269; Fax 631-231-8175
Web Site: http://www.novapublishers.com

Library of Congress Cataloging-in-Publication Data

ISBN: 978-1-62948-687-1

Published by Nova Science Publishers, Inc. † New York

CONTENTS

PREFACE

This book focuses on the geostationary weather satellites and polar weather satellites. Geostationary environmental satellites play a critical role in our nation's weather forecasting. These satellite, which are managed by the Department of Commerce's National Oceanic and Atmospheric Administration (NOAA), provide information on atmospheric, oceanic, climatic, and solar conditions that help meteorologists observe and predict regional and local weather events. They also provide a means of identifying the large-scale evolution of severe storms, such as forecasting a hurricane's path and intensity. The Editors of this book discuss the National Polar-orbiting Operational Environmental Satellite System (NPOESS) program, planned to be a state-of-the-art, environment-monitoring satellite system that would replace two existing polar-orbiting environmental satellite systems.

Chapter 1 – NOAA, with the aid of the National Aeronautics and Space Administration (NASA), is procuring the next generation of geostationary weather satellites. The GOES-R series is to replace the current series of satellites (called GOES-13, -14, and -15), which will likely begin to reach the end of their useful lives in 2015. This new series is considered critical to the United States' ability to maintain the continuity of satellite data required for weather forecasting through 2036.

GAO was asked to evaluate GOES-R. GAO's objectives were to (1) assess GOES-R progress and efforts to address key cost and schedule risks; (2) evaluate efforts to manage changes in requirements and whether any significant changes have recently occurred; and (3) evaluate the adequacy of GOES-R contingency plans. To do so, GAO analyzed program and contractor data, compared GOES-R schedules, requirements changes, and contingency plans to best practices by leading organizations, and interviewed officials at NOAA, NASA, and at other federal agencies that rely on GOES.

Chapter 2 – NOAA established the JPSS program in 2010 to replace aging polar satellites and provide critical environmental data used in forecasting weather and measuring variations in climate. However, program officials anticipate a gap in satellite data between the time that the S-NPP satellite reaches the end of its life and the JPSS-1 satellite becomes operational (see graphic). Given the criticality of satellite data to weather forecasts, the likelihood of a significant satellite data gap, and the potential impact of a gap on the health and safety of the U.S. population and economy, GAO added this issue to its High Risk List in 2013.

GAO was asked to review the JPSS program because of the importance of polar satellite data. GAO's objectives were to (1) evaluate NOAA's progress in sustaining the continuity of NOAA's polar-orbiting satellite system through S-NPP and JPSS satellites; (2) evaluate the quality of NOAA's program schedule; and (3) assess NOAA's plans to address potential gaps in polar satellite data. To do so, GAO analyzed program management status reports, milestone reviews, and schedule data; examined polar gap contingency plans; and interviewed agency and contractor officials.

Chapter 3 – Statement of Mary E. Kicza, NESDIS Assistant Administrator, National Oceanic and Atmospheric Administration.

Chapter 4 – Statement of Marcus A. Watkins, Director, Joint Agency Satellite Division, National Aeronautics and Space Administration.

In: Weather Satellites
Editor: Sherril G. Ryder

ISBN: 978-1-62948-687-1

Chapter 1

GEOSTATIONARY WEATHER SATELLITES: PROGRESS MADE, BUT WEAKNESSES IN SCHEDULING, CONTINGENCY PLANNING, AND COMMUNICATING WITH USERS NEED TO BE ADDRESSED*

United States Government Accountability Office

WHY GAO DID THIS STUDY

NOAA, with the aid of the National Aeronautics and Space Administration (NASA), is procuring the next generation of geostationary weather satellites. The GOES-R series is to replace the current series of satellites (called GOES-13, -14, and -15), which will likely begin to reach the end of their useful lives in 2015. This new series is considered critical to the United States' ability to maintain the continuity of satellite data required for weather forecasting through 2036.

GAO was asked to evaluate GOES-R. GAO's objectives were to (1) assess GOES-R progress and efforts to address key cost and schedule risks; (2) evaluate efforts to manage changes in requirements and whether any significant changes have recently occurred; and (3) evaluate the adequacy of

---

* This is an edited, reformatted and augmented version of United States Government Accountability Office, Publication No. GAO-13-597, dated September 2013.

GOES-R contingency plans. To do so, GAO analyzed program and contractor data, compared GOES-R schedules, requirements changes, and contingency plans to best practices by leading organizations, and interviewed officials at NOAA, NASA, and at other federal agencies that rely on GOES.

WHAT GAO RECOMMENDS

GAO is recommending that NOAA address weaknesses in managing reserves and scheduling, improve communications with satellite data users, and address shortfalls in contingency planning. NOAA concurred with GAO's recommendations and identified steps it is taking to implement them.

WHAT GAO FOUND

The National Oceanic and Atmospheric Administration (NOAA) has completed the design of its Geostationary Operational Environmental Satellite-R (GOES-R) series and made progress in building flight and ground components. While the program reports that it is on track to stay within its $10.9 billion life cycle cost estimate, it has not reported key information on reserve funds to senior management. Also, the program has delayed interim milestones, is experiencing technical issues, and continues to demonstrate weaknesses in the development of component schedules. These factors have the potential to affect the expected October 2015 launch date of the first GOES-R satellite, and program officials now acknowledge that the launch date may be delayed by 6 months. A launch delay would increase the time that NOAA is without an on-orbit backup satellite. It would also increase the potential for a gap in GOES satellite coverage should one of the two operational satellites (GOES-14 or -15) fail prematurely — a scenario given a 36 percent likelihood of occurring by an independent review team.

While the GOES-R program has established a process for managing requirements changes, it has not effectively involved key satellite data users. Since 2007, the GOES-R program decided not to develop 31 of the original set of GOES products and modified specifications on 20 remaining products. For example, NOAA decreased the accuracy requirement for the hurricane intensity product and decreased the timeliness of the lightning detection product.

However, key satellite data users were not fully informed about changes and did not have a chance to communicate their concerns about the impact of these changes on their operations. Until NOAA improves its communication with external satellite data users, obtains input from the users, and addresses user concerns when considering product changes, its changes could cause an unexpected impact on critical user operations.

NOAA has established contingency plans for the loss of its GOES satellites and ground systems that are generally in accordance with best practices; however, these plans are missing key elements. For example, NOAA did not work with the user community to address potential reductions in capability under contingency scenarios or identify alternative solutions for preventing a delay in the GOES-R launch date. Until NOAA addresses the shortfalls in its contingency plans and procedures, the plans may not work as intended in an emergency and satellite data users may not obtain the information they need to perform their missions.

ABBREVIATIONS

CDR	critical design review
FOR	flight operations review
GOES	Geostationary Operational Environmental Satellite
GOES-R	Geostationary Operational Environmental Satellite-R series
KDP	key decision point
MDR	mission definition review
MOR	mission operations review
NASA	National Aeronautics and Space Administration
NESDIS	National Environmental Satellite, Data, and Information Service
NOAA	National Oceanic and Atmospheric Administration
NSOF	NOAA Satellite Operations Facility
ORR	operational readiness review
PDR	preliminary design review
SDR	system definition review
SIR	system integration review

September 9, 2013

The Honorable Lamar S. Smith
Chairman

The Honorable Ralph Hall
Chairman Emeritus

The Honorable Eddie Bernice Johnson
Ranking Member
Committee on Science, Space, and Technology
House of Representatives

Geostationary environmental satellites play a critical role in our nation's weather forecasting. These satellites—which are managed by the Department of Commerce's National Oceanic and Atmospheric Administration (NOAA)—provide information on atmospheric, oceanic, climatic, and solar conditions that help meteorologists observe and predict regional and local weather events. They also provide a means of identifying the large-scale evolution of severe storms, such as forecasting a hurricane's path and intensity.

NOAA, through collaboration with the National Aeronautics and Space Administration (NASA), is procuring the next generation of geostationary weather satellites, called the Geostationary Operational Environmental Satellite–R (GOES-R) series. The GOES-R series consists of four satellites and is to replace the current series of geostationary environmental satellites as they reach the end of their useful lives. This new series is expected to provide the first major improvement in the technology of GOES instruments since 1994 and, as such, is considered critical to the United States' ability to maintain the continuity of data required for weather forecasting through the year 2036.

This report responds to your request that we review NOAA's GOES-R series program (GOES-R program). Specifically, our objectives were to (1) assess GOES-R progress and efforts to address key cost and schedule risks that we identified in our prior report, (2) evaluate efforts to manage changes in requirements and whether any significant changes have recently occurred, and (3) evaluate the adequacy of GOES-R contingency plans. To assess NOAA's progress in developing GOES-R and addressing key risks, we compared estimated and actual program deliverables and analyzed monthly program status briefings to identify current status and recent development challenges.

We also followed up on our prior concerns regarding reserve funds and scheduling practices by comparing the program's current level of reserve funding and two component schedules to best practices.[1] By recalculating reserve percentages based on supporting data and examining schedule anomalies through use of a standard template, we determined data in both areas to be reliable for the purposes of this audit. To assess NOAA's efforts to manage changes in requirements, we compared the agency's policies and practices to best practices identified by leading organizations[2] and identified major changes to the program over time. To evaluate the adequacy of the GOES-R contingency plan, we compared the GOES-R contingency plan to best practices in contingency planning identified by leading organizations.[3] We also interviewed program officials as well as key internal and external satellite data users.

We conducted this performance audit from October 2012 to September 2013 in accordance with generally accepted government auditing standards. Those standards require that we plan and perform the audit to obtain sufficient, appropriate evidence to provide a reasonable basis for our findings and conclusions based on our audit objectives. We believe that the evidence obtained provides a reasonable basis for our findings and conclusions based on our audit objectives. See appendix I for a complete description of our objectives, scope, and methodology.

BACKGROUND

Since the 1970s, geostationary satellites have been used by the United States to provide meteorological data for weather observation, research, and forecasting. NOAA's National Environmental Satellite, Data, and Information Service is responsible for managing the civilian operational geostationary satellite system, called GOES. Geostationary satellites can maintain a constant view of the earth from a high orbit of about 22,300 miles in space.

NOAA operates GOES as a two-satellite system that is primarily focused on the United States (see fig. 1). These satellites provide timely environmental data about the earth's atmosphere, surface, cloud cover, and the space environment to meteorologists and their audiences. They also observe the development of hazardous weather, such as hurricanes and severe thunderstorms, and track their movement and intensity to reduce or avoid major losses of property and life. The ability of the satellites to provide broad, continuously updated coverage of atmospheric conditions over land and oceans is important to NOAA's weather forecasting operations.

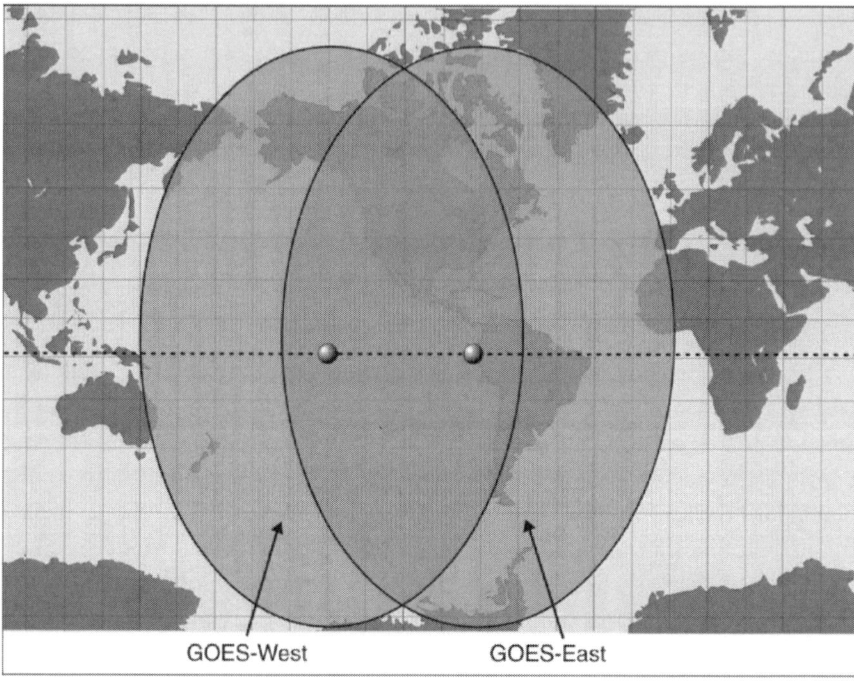

Source: NOAA (data), Mapart (map).

Figure 1. Approximate Geographic Coverage of the Geostationary Operational Environmental Satellites.

To provide continuous satellite coverage, NOAA acquires several satellites at a time as part of a series and launches new satellites every few years (see table 1). NOAA's policy is to have two operational satellites and one backup satellite in orbit at all times.

Four GOES satellites–GOES-12, GOES-13, GOES-14, and GOES-15– are currently in orbit. Both GOES-13 and GOES-15 are operational satellites with GOES-13 covering the eastern United States and GOES-15 in the western United States (see fig. 1). GOES-14 is currently in an on-orbit storage mode and available as a backup for the other two satellites should they experience any degradation in service. GOES-12 is at the end of its service life, but it is being used to provide limited coverage of South America. The GOES-R series is the next generation of satellites that NOAA is planning. The first two satellites in the series (called GOES-R and GOES-S) are planned for launch in October 2015 and February 2017, respectively.[4]

Table 1. Summary of the Procurement History of the Geostationary Operational Environmental Satellites

Series name	Procurement duration[a]	Satellites[b]
Original GOES[c]	1970-1987	1, 2, 3, 4, 5, 6, 7
GOES I-M	1985-2001	8, 9, 10, 11, 12
GOES N	1998-2010	13, 14, 15, Qd
GOES-R	2008-2024	R, S, T, U

Source: GAO analysis of NOAA data.

[a]Duration includes time from contract award to final satellite launch.

[b]Satellites in a series are identified by letters of the alphabet when they are on the ground (before launch) and by numbers once they are in orbit.

[c]The procurement of these satellites consisted of four separate contracts for (1) two early prototype satellites and GOES-1, (2) GOES-2 and -3, (3) GOES-4 through -6, and (4) GOES-G (failed on launch) and GOES-7.

[d]NOAA decided not to exercise the option for this satellite.

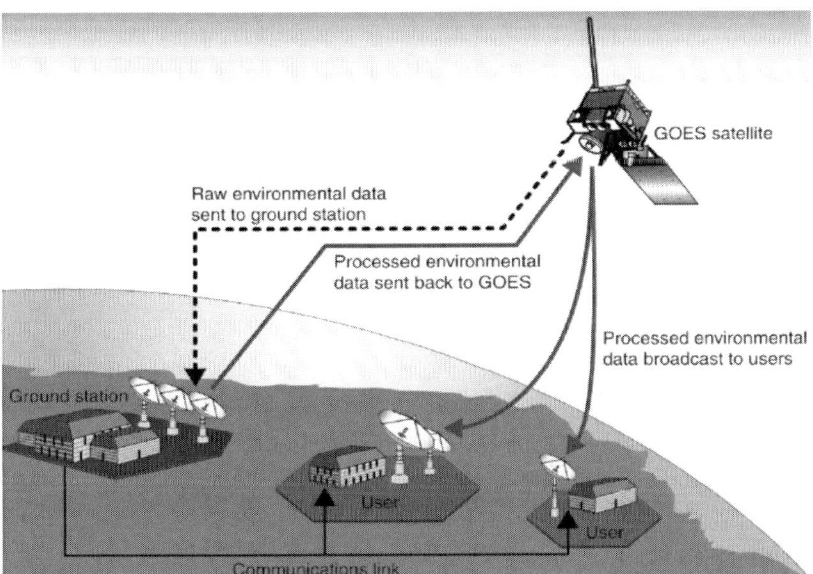

Source: GAO analysis of NOAA data.

Figure 2. Generic Data Relay Pattern for the Geostationary Operational Environmental Satellites.

Each of the operational geostationary satellites continuously transmits raw environmental data to NOAA ground stations. The data are processed at these ground stations and transmitted back to the satellite for broadcast to primary weather services and the global research community in the United States and abroad. Raw and processed data are also distributed to users via ground stations through other communication channels, such as dedicated private communication lines and the Internet. Figure 2 depicts a generic data relay pattern from a geostationary satellite to the ground stations and commercial terminals.

Overview of the GOES-R Program

NOAA established the GOES-R program to develop and launch the next series of geostationary satellites and to ensure the continuity of geostationary satellite observations. Since its inception, the GOES-R program has undergone several changes in cost and scope. As originally envisioned, GOES-R was to encompass four satellites hosting a variety of advanced technology instruments and providing 81 environmental products. The first two satellites in the series were expected to launch in September 2012 and April 2014. However, in September 2006, NOAA decided to reduce the scope and technical complexity of the GOES-R program because of expectations that total costs, which were originally estimated to be $6.2 billion, could reach $11.4 billion. Specifically, NOAA reduced the minimum number of satellites from four to two, cancelled plans for developing an advanced instrument (which reduced the number of planned satellite products from 81 to 68), and divided another instrument into two separate acquisitions. The agency estimated that the revised program would cost $7 billion and kept the planned launch dates unchanged.

Subsequently, NOAA made several other important decisions about the cost and scope of the GOES-R program. In May 2007, NOAA had an independent cost estimate completed for the GOES-R program. After reconciling the program office's cost estimate of $7 billion with the independent cost estimate of about $9 billion, the agency established a new program cost estimate of $7.67 billion. This was an increase of $670 million from the previous estimate. The program also moved the launch dates for the first two satellites to December 2014 and April 2016.

Table 2. Key Changes to the Geostationary Operational Environmental Satellite-R Series Program over Time

	August 2006 (baseline program)	September 2006	November 2007	February 2011
Number of satellites	4	2	2	4
Instruments or instrument changes	• Advanced Baseline Imager • Geostationary Lightning Mapper • Magnetometer • Space Environmental In-Situ Suite • Solar Imaging Suite (which included the Solar Ultraviolet Imager, and Extreme Ultraviolet/X-Ray Irradiance Sensor) • Hyperspectral Environmental Suite	• Advanced Baseline Imager • Geostationary Lightning Mapper • Magnetometer • Space Environmental In-Situ Suite • Solar Ultraviolet Imager • Extreme Ultraviolet/X-Ray Irradiance Sensor	No change	No change
Number of satellite products	81	68	34 baseline 34 optional	34 baseline 31 optional
Life cycle cost estimate (in then- year dollars)	$6.2 billion—$11.4 billion (through 2034)	$7 billion (through 2028)	$7.67 billion (through 2028)	$10.9 billion (through 2036)a
Estimated launch dates for GOES-R and S	GOES-R: September 2012 GOES-S: April 2014	GOES-R: September 2012 GOES-S: April 2014	GOES-R: December 2014 GOES-S: April 2016	GOES-R: October 2015 GOES-S: February 2017

Source: GAO analysis of NOAA data.

[a]Based on NOAA's fiscal year 2012 budget baseline, $7.64 billion of this cost estimate was for the first two satellites in the series, GOES-R and GOES-S. The cost for the remaining two satellites— GOES-T and GOES-U—was estimated at $3.22 billion.

Further, in November 2007, to mitigate the risk that costs would rise, program officials decided to remove selected program requirements from the baseline program and treat them as contract options that could be exercised if funds allowed. These requirements included the number of products to be distributed, the time to deliver the remaining products (product latency), and how often these products would be updated with new satellite data (refresh rate). For example, program officials eliminated the requirement to develop and distribute 34 of the 68 envisioned products, including low cloud and fog, sulfur dioxide detection, and cloud liquid water. Program officials included the restoration of the requirements for the products, latency times, and refresh rates as options in the ground system contract that could be acquired at a later time.

Program officials later reduced the number of products that could be restored as a contract option (called option 2) from 34 to 31 because they determined that two products were no longer feasible and two others could be combined into a single product.

In late 2009, NOAA changed the launch dates for the first two satellites to October 2015 and February 2017, in part due to a bid protest related to award of the spacecraft contract. More recently, NOAA restored two satellites to the program's baseline, making GOES-R a four-satellite program once again. In February 2011, as part of its fiscal year 2012 budget request, NOAA requested funding to begin development for two additional satellites in the GOES-R series—GOES-T and GOES-U.

The program estimates that the development for all four satellites in the GOES-R series—GOES-R, GOES-S, GOES-T, and GOES-U—is to cost $10.9 billion through 2036, an increase of $3.2 billion over its prior life cycle cost estimate of $7.67 billion for the two-satellite program. See table 2 for an overview of key changes to the GOES-R program.

Program and Program Office Structure

The GOES-R program is divided into flight and ground projects that have separate areas of responsibility and oversee different sets of contracts. The flight project, which is managed by NASA, includes instruments, spacecraft, launch services, satellite integration, and on-orbit satellite initialization. Table 3 summarizes the GOES-R instruments and their planned capabilities.

The ground project is directed by NOAA and is made up of three main components: the core ground system, an infrastructure of antennas, and a product access subsystem. In turn, the core ground system comprises four functional modules supporting operations, product generation, product distribution, and configuration control. Key components of the ground project are described in table 4.

Table 3. Geostationary Operational Environmental Satellite-R Series Instruments

Planned instrument	Description
Advanced Baseline Imager	Expected to provide variable area imagery and radiometric information of the earth's surface, atmosphere, and cloud cover. Key features include • monitoring and tracking severe weather; • providing images of clouds to support forecasts; and • providing higher resolution, faster coverage, and broader coverage simultaneously.
Geostationary Lightning Mapper	Expected to continuously monitor total lightning (in-cloud and cloud-to-ground) activity over the United States and adjacent oceans and to provide a more complete dataset than previously possible. Key features include • detecting lightning activity as an indicator of severe storms and convective weather hazard impacts to aviation; and • providing a new capability to GOES for long-term mapping of total lightning that only previously existed on NASA low-earth-orbiting research satellites.
Magnetometer	Expected to provide information on the general level of geomagnetic activity, monitor current systems in space, and permit detection of magnetopause crossings, sudden storm commencements, and substorms.
Space Environmental In-Situ Suite	Expected to provide information on space weather to aid in the prediction of particle precipitation, which causes disturbance and disruption of radio communications and navigation systems. Key features include • measuring magnetic fields and charged particles; • providing improved heavy ion detection, adding low-energy electrons and protons; and • enabling early warnings for satellite and power grid operation, telecom services, astronauts, and airlines.
Solar Ultraviolet Imager	Expected to provide coverage of the entire dynamic range of solar X-ray features, from coronal holes to X-class flares, and will provide quantitative estimates of the physical conditions in the Sun's atmosphere. Key features include • providing information used for geomagnetic storm forecasts, and power grid performance; and • providing observations of solar energetic particle events related to flares.
Extreme Ultraviolet/X-Ray Irradiance Sensor	Expected to detect solar soft X-ray irradiance and solar extreme ultraviolet spectral irradiance. Key features include • monitoring solar flares that can disrupt communications and degrade navigational accuracy, affecting satellites, astronauts, high latitude airline passengers; and • monitoring solar variations that directly affect satellite drag/tracking and ionospheric changes, which impact communications and navigation operations.

Source: GAO analysis of NOAA data.

Table 4. Key Components of the Geostationary Operational Environmental Satellite-R Series Ground Project

Component	Description
Core Ground System	Expected to (1) provide command of operational functions of the spacecraft and instruments, (2) receive and process information from the instruments and spacecraft, (3) distribute satellite data products to users, and (4) provide configuration control and a common infrastructure and set of services for the satellite and instruments.
Antennas	Expected to provide six new antenna stations and modify four existing antennas to receive GOES-R data. The antenna contract is also expected to include the construction of related infrastructure, software development for control systems, and maintenance.
Product Distribution and Access System	Expected to provide ingestion of data and distribution for GOES-R products and data to authorized users. When completed, this system will be integrated into the core ground system.

Source: GAO analysis of NOAA data.

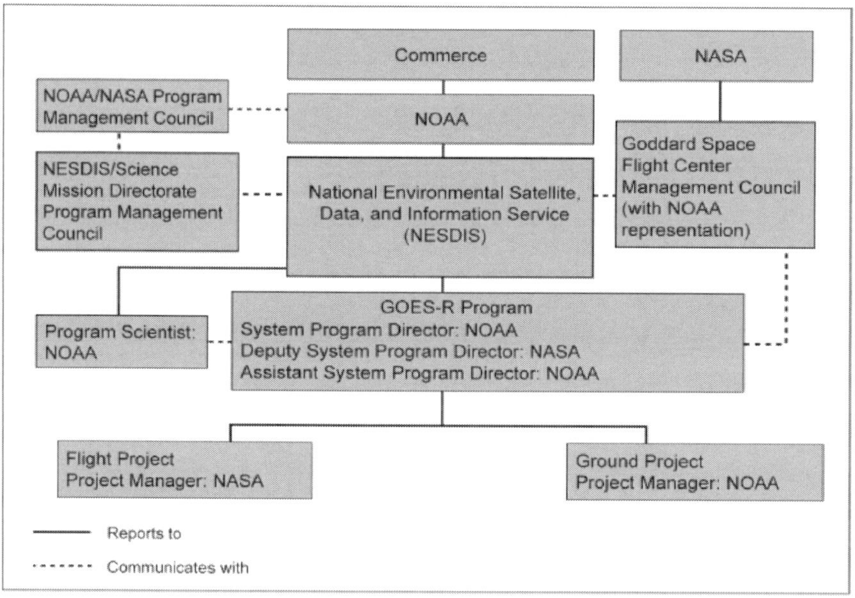

Source: NOAA.

Figure 3. Organizational Structure and Staffing of the Geostationary Operational Environmental Satellite-R Series Program.

NOAA is responsible for GOES-R program funding and overall mission success. The NOAA Program Management Council, which is chaired by NOAA's Deputy Undersecretary, is the oversight body for the GOES-R program. However, since it relies on NASA's acquisition experience and technical expertise to help ensure the success of its programs, NOAA implemented an integrated program management structure with NASA for the GOES-R program (see fig. 3). NOAA also located the program office at NASA's Goddard Space Flight Center.

Prior Reports Made Recommendations to Address Program Weaknesses

In recent years, we issued a series of reports aimed at addressing weaknesses in the GOES-R program.[5] Key areas of focus included (1) improving communications with external data users, (2) developing contingency plans, and (3) addressing key cost and schedule risks.

- **Improving communications with external users.** In September 2010, we reported that while NOAA had identified GOES data users and involved internal NOAA users in developing and prioritizing GOES-R requirements, it had not adequately involved other federal users who rely on GOES data by documenting their input and communicating major changes to the program.[6] We recommended that the program establish processes for satellite data requirements definition and prioritization that include documented input from external users, as well as processes to notify these non-NOAA agencies of GOES-R program status and changes. In February 2012, the GOES-R program developed a communications plan that described how external stakeholders would be notified of GOES-R progress, status changes, and other relevant activities. However, NOAA has not yet fully implemented the plan, as demonstrated by the communication shortfalls discussed later in this report.
- **Developing contingency plans.** In September 2010, we reported that while there was a potential gap in backup coverage due to satellite launch delays, NOAA had not established adequate continuity plans for its geostationary satellites.[7] We recommended that the program's plan include implementation procedures, resources, staff roles, and timetables needed to transition to a single satellite, a foreign satellite,

or other solution. In December 2012, NOAA finalized a contingency plan that generally included these elements. However, more work remains to ensure that the plan is viable.

More recently, in February 2013, we added the potential gaps in weather satellite data to our biennial High-Risk list.[8] In that report, we noted that NOAA had established a contingency plan for a potential gap in the GOES program, but it needed to demonstrate its progress in coordinating with the user community to determine their most critical requirements, conducting training and simulations for contingency operations scenarios, evaluating the status of viable foreign satellites, and working with the user community to account for differences in product coverage under contingency operations scenarios. We also stated that NOAA should update its contingency plan to provide more details on its contingency scenarios, associated time frames, and any preventative actions it is taking to minimize the possibility of a gap.

- **Addressing key cost and schedule risks.** In June 2012, we reported that the GOES-R program might not be able to ensure that it had adequate resources to cover unexpected problems in remaining development, and that unresolved schedule deficiencies existed in its integrated master schedule and contractor schedules. We also reported that the program estimated a 48 percent chance that the planned GOES-R launch date of October 2015 would be reached.[9] We recommended that the program assess and report on the reserves needed for completing remaining development for each satellite in the series, and address shortfalls in the schedule management practices we identified such as eliminating unnecessary constraints and creating a realistic allocation of resources, in order to minimize the likelihood of a potential gap. The agency agreed with these recommendations and took steps to address them by identifying needed reserve levels and refining program schedules.

NOAA Has Made Progress in Developing GOES-R, but Continues to Face Challenges that Could Increase the Risk of a Satellite Data Gap

NOAA has completed its design of the GOES-R program, and has made progress in building components of the flight and ground segments. Program officials also report that the program is operating within its estimated budget of $10.9 billion. However, key information on reserves has not been reported to management. Further, both the flight and ground segments have experienced delays in achieving major milestones due to technical challenges, and weaknesses in the development of master schedules could cause further delays. Program officials stated that they have made improvements on how they manage cost reserves and schedules, but acknowledged that there will always be opportunities for improvement because the reserves and schedules are so dynamic on a big program like GOES-R. These challenges have the potential to impact the expected launch date of the first GOES-R satellite, which would delay the availability of an on-orbit backup and increase the potential for a gap in GOES satellite coverage should either of the two operational satellites fail prematurely.

Program Has Completed Design and Begun Building Components of the First Satellite

NASA and NOAA are following NASA's standard space system life cycle on the GOES-R program. This life cycle includes distinct phases, including concept and technology development; preliminary design and technology completion; final design and fabrication; system assembly, integration and testing, launch and checkout; and operations and sustainment. There are key program reviews throughout each of the phases, including preliminary design review, critical design review, and system integration review. NOAA and NASA jointly conduct key reviews on the flight and ground segments individually as well as for the program as a whole, and then make decisions on whether to proceed to the next phase. Figure 4 provides an overview of the life cycle phases, key program reviews, and associated decision milestones. In addition, the key reviews are described in table 5.

The GOES-R program has completed final design and begun building components of the flight and ground systems. Specifically, the program completed critical design reviews for the flight and ground projects and for the overall program between April and November 2012. In its evaluation of the program as part of the critical design review, an independent review board complimented the program on several recent achievements, stating that the program was beyond the level of maturity expected at that phase, and that the program's planning was a major factor in the launch date of the first satellite remaining October 2015.

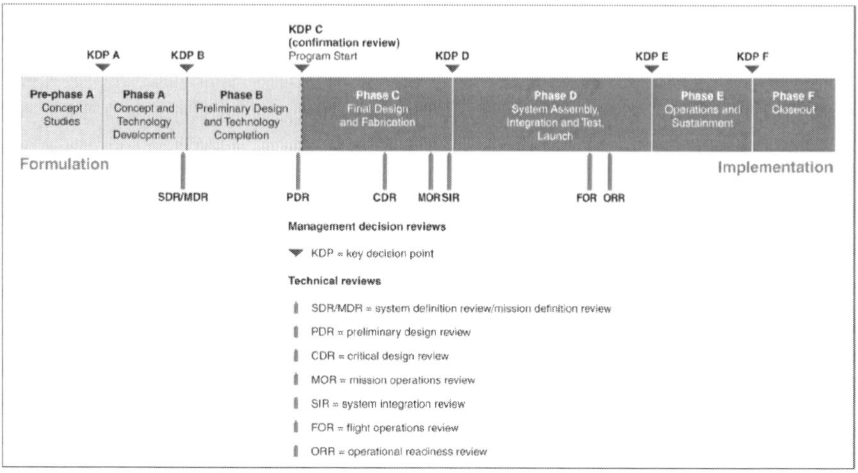

Source: NASA data and GAO analysis.

Note: According to a NASA official, the MOR and FOR are considered lower-level reviews and are not mandated by NASA's primary procedural requirements. They are, however, key mission reviews required by NASA's Goddard Space Flight Center.

Figure 4. NASA's Life Cycle for Flight Systems.

Table 5. Major Development Reviews for the Geostationary Operational Environmental Satellite-R Series

Review	Description
System Definition Review	Performed on the flight and ground segments individually, and then on the program as a whole, this review is to examine the proposed system architecture/design and demonstrate that a system that fulfills the mission objectives can be built within existing constraints.
Preliminary Design Review	Performed on the flight and ground segments individually, and then on the program as a whole, this review is to demonstrate that the preliminary design meets all system requirements with acceptable risk and within the cost and schedule constraints and to establish the basis for proceeding with detailed design.
Critical Design Review	Performed on the flight and ground segments individually, and then on the program as a whole, this review is to evaluate the completed detailed design of the element and subsystem products insufficient detail to provide approval for a production stage.

Review	Description
Mission Operations Review	Performed programwide, this review is to establish the adequacy of plans and schedules for ground systems and flight operations preparation, and to justify readiness to proceed with implementation of the remaining required activities. It is typically held subsequent to completion of detail design and fabrication activity, but prior to initiation of major integration activities of flight or ground-system elements.
System Integration Review	Performed programwide, this review is to evaluate the readiness of the project to start system assembly, test, and launch operations. The objectives of the review include ensuring that planning is adequate for all remaining system activities and that available cost and schedule resources support completion of all necessary remaining activities with adequate margin.
Flight Operations Review	This review is to present the results of mission operations activities and show that the program has verified compliance with all requirements and demonstrated the ability to execute all phases and modes of mission operations, data processing, and analysis.
Operational Readiness Review	This review is to examine characteristics and procedures used in the system's operation and ensures that all system and support hardware, software, personnel, and procedures are ready for operations and that user documentation accurately reflects the deployed state of the system. It is typically held near the completion of pre-launch testing between the flight segment and the ground system.

Source: GAO analysis of NOAA documentation.

As the spacecraft and instruments are developed, NASA conducts several interim reviews and tests before proceeding to the next major program-level review, the system integration review. These include a pre-environmental review, which represents the conclusion of an initial round of testing before exposing the instrument to testing under adverse environmental conditions; environmental testing of key functions under adverse conditions; and a pre-shipment review, which is conducted on each instrument to ensure it is ready to be shipped for integration and testing on the spacecraft.

The GOES-R flight components are in various stages leading up to the system integration review. Of the six GOES-R instruments, one has completed environmental testing and its pre-shipment review; four instruments are in the midst of these reviews and tests; and one instrument has not yet passed its pre-environmental review. In addition, the program began building the spacecraft in February 2013. Table 6 provides more information on progress made on the key flight project components.

Table 6. Development Status of Flight Project Components for the Geostationary Operational Environmental Satellite-R Satellite, as of August 2013

Key component	Recent progress
Advanced Baseline Imager	• Pre-environmental review completed in November 2012 • Environmental testing completed
Extreme Ultraviolet/X-Ray Irradiance Sensor	• Instrument fully assembled and tested • Pre-environmental review conducted in July 2012 • Pre-ship review conducted in April 2013
Geostationary Lightning Mapper	• Assembly of some subcomponents completed, others continuing • Subcomponent testing is under way
Magnetometer	• Selected components have completed readiness reviews and tests • Environmental testing is under way
Solar Ultraviolet Imager	• Pre-environmental review completed in November 2012 • Environmental testing is under way • Pre-ship review scheduled for October 2013
Space Environmental In-Situ Suite	• Individual component testing completed • Pre-environmental review conducted in May 2013
Spacecraft	• Corestructure testing completed; multiple components delivered • Integration of subsystems under way • Construction of the system module that will host instruments is under way

Source: GAO analysis of NOAA documentation.

Similar to the flight project, major ground system milestones are focused on building and testing components and the program has made progress in this area. Specifically, on the core ground system, a prototype for the operations module was delivered in late 2012 and used for initial testing and training.[10] In July 2013, the ground project delivered the iteration of the operations module that will be used to support the first satellite. In addition, the program has installed antenna dishes at NOAA's primary satellite communications site, and completed two key reviews of antennas at the GOES remote backup site. The Product Distribution and Access System recently completed a review that will allow testing to begin on its first release. An integration review for ground components is also expected to take place in January 2014. More detail on the progress of the ground project can be seen in table 7.

Table 7. Development Status of the Geostationary Operational Environmental Satellite-R Series Ground Project Components, as of August 2013

Key component	Recent progress
Core Ground Segment	• Critical design review completed in April2012 • Completed readiness review for receipt of GOES-R antennas • Delivery and installation of a prototype of the mission operations module completed; next iteration of the module delivered in July 2013
Antenna System	• Contractor demonstrated ability to produce 8 of 13 components; remainder due by the end of2013 • Installation of the first two antenna structures has been completed; the third antenna structure is scheduled to be completed in fiscal year 2014 • Supporting infrastructure built, and two key reviews completed, for remote back-up antenna site
Product Distribution and Access System	• Testing begun on first increment/release

Source: GAO analysis of NOAA documentation.

The program's next major milestone is a programwide system integration review, which is scheduled for March 2014. Based on the results of that review, NOAA and NASA will decide whether to move the program to the next phase: the system assembly, integration and test, and launch and checkout phase.

Contingency Reserves Are Generally in Line with Goals for Overall Program Development; Reporting on Reserve Values Remains Limited

The GOES-R program is estimated to cost $10.9 billion. As of February 2013, the program estimated that this amount was divided into four categories, with $6.0 billion for the flight project, $1.7 billion for the ground project, $2.0 billion for other program costs (including, among other things, program/project support and formulation) and $1.2 billion for operations and support. Program officials reported that the program is currently operating without cost overruns on any of its main components, but noted that the program life cycle costs may increase by $150 to $300 million if full funding in the current fiscal year is not received.

A portion of the amounts planned for the flight project and ground project are allocated to contingency reserves (also called management reserve). The program also keeps a programwide contingency allocation separate from those of the flight and ground projects.

A contingency reserve provides program managers ready access to funding in order to resolve problems as they occur and may be necessary to cover increased costs resulting from unexpected design complexity, incomplete requirements, or other uncertainties.[11] NASA's Goddard Space Flight Center requires its flight projects, including GOES-R, to maintain contingency reserves during system development, at a level of 25 percent of development costs.[12]

The GOES-R program requires its flight and ground projects to maintain 20 percent of planned remaining development costs as reserve funding.[13] The program office also maintains contingency reserves equal to 10 percent of planned remaining development costs to cover program support costs and to supplement the flight and ground projects' reserves if necessary.

According to a NOAA program official, the GOES-R program is able to meet NASA's requirement through the combination of the 20 percent flight and ground project requirements and the supplemental 10 percent program-level reserve. An official also stated that the method of keeping separate reserves at the program and project levels was chosen as it was successful on past projects.

The GOES-R flight project, ground project, and program office are at or above the amount of reserves they are required to carry. Specifically, as of March 2013, the overall contingency reserve percentages for the flight and ground projects were at 20 and 28 percent, respectively, which are at or above the required level of 20 percent.

The program reserves were at 11 percent, slightly above the required level of 10 percent. Reserve values and percentages are provided in table 8.

Table 8. Reserve Levels for the Geostationary Operational Environmental Satellite-R Series Program, as of March 2013

Component	Required reserve	Remaining reserve[b]
Flight	20%	20%
Ground	20%	28%
Program	10%	11%
Total	25%[a]	29%

Source: NOAA data and GAO analysis of NOAA data.

Notes:

[a] NASA and NOAA officials stated that the allocation of reserves among flight, ground, and program components meets NASA's requirement of 25 percent because the program category includes reserve funds that can be used to supplement the flight and ground components.

[b] A series of adjustments are made to the flight, ground, and program budget amounts before the reserve percentage is calculated; thus the reserve percentage cannot be reached by dividing contingency reserves by total budget authority.

Reporting on Reserves Is Not Sufficiently Detailed or Transparent

We previously reported that, in order to oversee GOES-R contingency funding, senior managers should have insight into the amount of reserves set aside for each satellite in the program and detailed information on how reserves are being used on both the flight and ground components. While the GOES-R program continues to regularly report on contingency funds, it does not report key information on the status of its reserve funding to senior level management.[14]

At monthly program management council meetings, the program reports summary information, such as the total value of contingency reserves and reserve percentage held for each fiscal year. Reserve totals are given for the flight and ground projects as well as for the program overall. However, the program does not report on the reserves needed for completing remaining development for each satellite in the series or provide detailed information on how reserves are being used. Thus, for example, if the later satellites in the series have a high level of reserves, they could mask low reserve values for earlier satellites in the series. Further, in its monthly presentations to senior managers and NOAA executives, the program does not include information on the cause of any changes in reserve values from the prior month or the assumptions it makes when calculating reserves. For example, the flight reserve value recently went up by 2 percentage points because the program decided to include reserve funding for the GOES-T satellite in 2018, and the ground reserve values went down by 10 percentage points because the program shifted reserve funding from the ground to the flight projects. Neither of these changes was identified or explained in the monthly presentations. The lack of insight on how the reserves are calculated and modified could lead executives to misinterpret the size of the remaining reserves. Program officials noted that they took steps after our previous report to clarify what they report about reserves, but noted that the amount of information needed to fully explain reserve calculations and changes could be too much information for an executive-level briefing. Without regularly providing sufficiently detailed budget information, it may be more difficult for program management to have the information they need to make the best decisions possible regarding the program's future funding.

Table 9. Delays in Milestones for the Geostationary Operational Environmental Satellite-R Series Program

Program milestone	Date planned (as of Apr 2012)	Date completed or planned (as of Mar 2013)	Delay
Mission operations review	January 2013	January 2014	12 months[a]
End-to-end test #1	February 2014	May 2014	3 months
End-to-end test #2	May 2014	August 2014	3 months
Program milestone	Date planned (as of Apr 2012)	Date completed or planned (as of Mar 2013)	Delay
End-to-end test #3	August 2014	December 2014	4 months
Flight operations review	September 2014	January 2015	4 months
End-to-end test #4	December 2014	March 2015	3 months
End-to-end test #5	July 2015	July 2015	No change

Source: GAO analysis of NOAA data.

[a]Program officials stated that they had erroneously scheduled the mission operations review too soon, and moved the date by 9 months to better reflect when the review was needed. Therefore, only 3 of the 12 months were attributable to a delay.

Recent and Potential Milestone Delays and Continued Weaknesses in Scheduling Practices Increase the Potential for a Delayed Launch

The GOES-R program established programwide milestones, including the mission operations review and flight operations review, to determine the program's ability to proceed to system integration and to complete mission operations, respectively. It also established five end-to-end system tests to validate compatibility between the space and ground segments before the launch of the first satellite.

However, over the past year, the program delayed many of these key milestones and tests. Delays in the mission operations review means that the large-scale integration of flight and ground components will not occur until 21 months prior to launch. Similarly, delaying end-to-end tests until 17 months prior to launch will allow the program less time to respond to any problems that occur. Table 9 highlights key milestones and the extent of recent delays.

Continued Technical Issues Could Cause Further Delays

The GOES-R program is also experiencing technical issues on the flight and ground projects that could cause further schedule delays.

- The original supplier for a key component on the spacecraft moved to a different facility, introducing risk due to the loss of experienced personnel and the impact on schedule. This led the program to find an alternative supplier. While a design review was performed to confirm resolution of the issue in April 2013, this change may lead to a delay of up to 6 months in integrating the component on the spacecraft. Program officials noted that this delay is not expected to impact the program's critical path or major milestones.
- The Geostationary Lightning Mapper's electronics unit experienced problems during testing, which led the program office to delay the tests.[15] The program is considering several options to address this issue, including using the electronics unit being developed for a later GOES satellite to allow key components to proceed with testing. If the issue cannot be resolved, it would affect the instrument's performance. As a result, the program is also considering excluding the Geostationary Lightning Mapper from the first GOES satellite. The program plans to make its decision on whether or not to include the instrument in late 2013. The removal of this instrument would cause a significant reduction in the satellite's functionality. Key GOES users have stated that they would prefer that NOAA delay launching the GOES-R satellite rather than launch it without the Geostationary Lightning Mapper.
- The program delayed the start of work on the ground system at the NOAA satellite operations facility by three months, from a planned date of October 2012 to January 2013, following a bid protest of the

award of a contract to upgrade the facility. This delay compressed an already tight schedule for testing the ground system.

- Testing for a number of ground system requirements has been postponed until future releases and builds, potentially causing modification to the schedule for these future products.
- Power amplifiers for the antenna systems experienced higher than expected failure rates, which could lead to schedule delays and decreases in operational availability.

Given that fewer than 3 years remain before GOES-R's expected launch in October 2015, continued delays in key milestones and reviews decrease the likelihood that the launch date will be met. Program officials recently acknowledged that the GOES-R launch date may be delayed by about 6 months, and attributed the cause of the delay to a shortfall of $54 million in anticipated funding in fiscal year 2013.[16]

Scheduling Practices Improved, but Weaknesses Remain

Delays in the program's remaining schedule are also at risk of further growth due to weaknesses in the program's scheduling methods. Program schedules not only provide a road map for systematic program execution, but also provide the means by which to gauge progress, identify and address potential problems, and promote accountability. Achieving success in managing large-scale programs depends in part on having an integrated and reliable schedule that defines, among other things, when work activities and milestone events will occur, how long they will take, and how they are related to one another. Without such a reliable schedule, program milestones may slip.

In June 2012, we reported on weaknesses in program schedules that comprised portions of the program's Integrated Master Schedule, including subordinate schedules for the spacecraft and core ground system. At that time, our work identified nine best practices associated with developing and maintaining a reliable schedule.[17] These are (1) capturing all activities, (2) sequencing all activities, (3) assigning resources to all activities, (4) establishing the duration of all activities, (5) integrating schedule activities horizontally and vertically, (6) establishing the critical path for all activities, (7) identifying reasonable float time between activities, (8) conducting a schedule risk analysis, and (9) updating the schedule using logic and durations. See table 10 for a description of each of these best practices.

Table 10. Description of Scheduling Best Practices

Practice	Description
Capturing all activities	The schedule should reflect all activities (steps, events, outcomes, etc.) as defined in the program's work breakdown structure to include activities to be performed by both the government and its contractors.
Sequencing all activities	The schedule should sequence activities in the order that they are to be implemented. In particular, activities that must finish prior to the start of other activities (i.e., predecessor activities), as well as activities that cannot begin until other activities have been completed (i.e., successor activities) should be identified.
Assigning resources to all activities	The schedule should reflect who will do the work activities, whether all required resources will be available when they are needed, and whether there are any funding or time constraints.
Establishing the duration of all activities	The schedule should reflect the duration of each activity. These durations should be as short as possible and have specific start and end dates.
Integrating schedule activities horizontally and vertically	The schedule should be horizontally integrated, meaning that it should link the products and outcomes associated with sequenced activities. The schedule should also be vertically integrated, meaning that there is traceability among varying levels of activities and supporting tasks and subtasks.
Establishing the critical path for all activities	The critical path represents the chain of dependent activities with the longest total duration in the schedule.
Identifying reasonable float time between activities	The schedule should identify a reasonable amount of float—the time that an activity can slip before the delay affects the finish milestone—so that schedule flexibility can be determined. As a general rule, activities along the critical path typically have the least amount of float.
Conducting a schedule risk analysis	A schedule risk analysis is used to predict the level of confidence in the schedule, determine the amount of time contingency needed, and identify high-priority schedule risks.
Updating the schedule using logic and durations to determine the dates	The schedule should use logic and durations in order to reflect realistic start and completion dates, be continually monitored to determine differences between forecasted completion dates and planned dates, and avoid logic overrides and artificial constraint dates.

Source: GAO analysis of government and industry practices in GAO-09-3SP.

In a previous report, we observed that important schedule components in GOES-R related schedules had not been included or completed, and recommended that these shortfalls be addressed.[18]

Table 11. Assessment of Selected Schedules Use of Best Practices over Time

Scheduling best practice	Spacecraft schedules		Core ground schedules	
	July 2011	November 2012	July 2011	November 2012
Best practice 1: Capturing all activities	◕	●	◑	◕
Best practice 2: Sequencing all activities	◑	◑	◑	◑
Best practice 3: Assigning resources to all activities	◔	◑	◑	◔
Best practice 4: Establishing the duration of all activities	◕	◕	◕	◕
Best practice 5: Integrating schedule activities horizontally and vertically	◕	◑	◑	◔
Best practice 6: Establishing the critical path for all activities	◕	◕	◔	◑
Best practice 7: Identifying float on activities and paths	◑	◑	◔	◔
Best practice 8: Conducting a schedule risk analysis	◔	◔	◔	◑
Best practice 9: Updating the schedule using logic and durations to determine the dates	●	◕	◕	●

Source: GAO analysis of schedules provided by GOES-R, documents and information received from GOES-R officials. Key

● The agency/contractor has fully met the criteria for this best practice

◕ The agency/contractor has substantially met the criteria for this best practice

◑ The agency/contractor has partially met the criteria for this best practice

◔ The agency/contractor has minimally met the criteria for this best practice

○ The agency/contractor has not met the criteria for this best practice

NOAA has since improved selected practices on its spacecraft and core ground schedules, but other practices stayed the same or worsened. Specifically, for the spacecraft, 2 practices were improved, 5 stayed the same, and 2 became weaker. For the core ground system, 4 practices were improved, 3 stayed the same, and 2 became weaker. Table 11 compares our assessments of the spacecraft and core ground system schedules in July 2011 and November 2012.

NOAA has improved elements of the schedules for both components. Specifically, the spacecraft schedule has eliminated level of effort activities[19] and has assigned resources for a greater percentage of activities. The core ground schedule now has an automated process by which all subcontractor records are combined to create an integrated schedule. It has a series of connected activities that lead to what contractor officials consider its main milestone delivery, and has implemented a detailed schedule risk analysis for a key upcoming release.

However, scheduling issues remain on the schedules for both components. For example, both schedules have issues with sequencing remaining activities and integration between activities. Regarding the spacecraft schedule, there is a small subset of activities with incomplete links between activities, and more than 20 percent of remaining detail activities have lags, or a set number of days between an activity and its successor. In the core ground schedule, a number of activities are missing either predecessor or successor activities, and there are several activities representing the end of the project on or about the same date. Without the right linkages, activities that slip early in the schedule do not transmit delays to activities that should depend on them. When this happens, the schedule will not provide a sufficient basis for understanding the program as a whole, and users of the schedule will lack confidence in the dates and the critical path.

Both schedules also have a very high average of total float time for detailed activities.[20] Specifically, total float time is greater than two months for nearly two-thirds of remaining detailed activities in the spacecraft schedule, and at least a year for more than 10 percent of remaining detail activities in the core ground schedule. In the case of spacecraft, officials stated that high levels of float time were often due to activities that had been completed at one time for several satellites, only one of which was immediately needed. Officials also provided detailed information on the activities with the highest amount of float. In the case of the core ground schedule, officials stated that many activities occurring after the main milestone date, which occurs nearly five years prior to the end of the schedule, do not have a true successor, and therefore are calculated only to the end of the contract. Officials also stated that values and trends in float time are monitored regularly for both schedules. Such high values of total float time can falsely depict true project status, making it difficult to determine which activities drive key milestone dates. Without reasonable values of total float time, it cannot be used to identify activities that could be permitted to slip and thus release and reallocate resources to activities that require more resources to be completed on time.

In addition, the project's critical path does not match up with activities that make up the driving path[21] on the core ground schedule. Contractors monitor a driving path monthly to both major and minor milestone deliveries. However, until the schedule can produce a true critical path, it will be more difficult for the program office to provide reliable time line estimates or identify when problems or changes may occur and their effect on downstream work. Also, without a valid critical path to the end of the schedule, management cannot focus on activities that will have a detrimental effect on the key project milestones and deliveries if they slip.

Further, neither schedule file has fully integrated resources with schedule activities. As of November 2012, contractor officials stated that the ground system schedule was not feasible given available resources and that they were in the process of revising their immediate schedules to make them feasible. The spacecraft schedule contains major resource categories that correspond to contractor sites and work phases. However, thresholds for overruns of resource allocations are functionally disabled within the schedules through the setting of an arbitrarily high value for maximum resources per category. In response, contractor officials stated that account managers are responsible for monitoring resource levels and that weekly meetings are held to ensure that resource issues are discussed. Information on resource needs and availability in each work period assists the program office in forecasting the likelihood that activities will be completed as scheduled. If the current schedule does not allow insight into current or project allocation of resources, then the risk of delays in the program's schedule is significantly increased.

Deficiencies in scheduling practices such as the ones outlined here could increase the likelihood of launch date delays, because decision making would be based on data that does not accurately depict current status, thus impeding management's ability to conduct meaningful oversight on the program's schedules. Program officials noted that they have made improvements in scheduling practices, but explained that because the schedules are so dynamic there are always areas for improvement. Lack of the proper understanding of current program status due to schedules that are not fully reliable undercuts the ability of the program office to manage a high-risk program like GOES-R.

Delays in the GOES-R Launch Date Could Increase the Risk of a Satellite Data Gap

Potential delays in the launch date of the first GOES-R satellite would increase the risk of a gap in GOES satellite coverage. NOAA's policy is to have two operational satellites and one backup satellite in orbit at all times. This policy proved useful in December 2008 and again in September 2012, when the agency experienced problems with one of its operational satellites, but was able to move its backup satellite into place until the problems had been resolved.

NOAA is facing a period of at least a year when it will not have a backup satellite in orbit. Specifically, in April 2015, NOAA expects to retire one of its operational satellites (GOES-13) and move its backup satellite (GOES-14) into operation. Thus, the agency will have only two operational satellites in orbit—and no backup satellite—until GOES-R is launched and completes an estimated 6-month post-launch test period. If GOES-R is launched in October 2015, the soonest it could be available for operational use would be April 2016. Any delay to the GOES-R launch would extend the time without a backup to more than one year. Figure 5 shows anticipated operational and test periods for the two most recent series of GOES satellites.

In addition to the year or more during which no back-up satellite would be available, there is a chance that NOAA would have to operate with a single operational satellite. In December 2012, an independent review board estimated that there is a 36 percent chance that the GOES constellation would have only one operational satellite at the expected date of GOES-R's launch. Thus, if NOAA were to experience a problem with either of its operational satellites before GOES-R is in orbit and operational, it would need to rely on older satellites that are beyond their expected operational lives and may not be fully functional. Without a full complement of operational GOES satellites, the nation's ability to maintain the continuity of data required for effective weather forecasting could be compromised. This, in turn, could put the public, property, and the economy at risk.

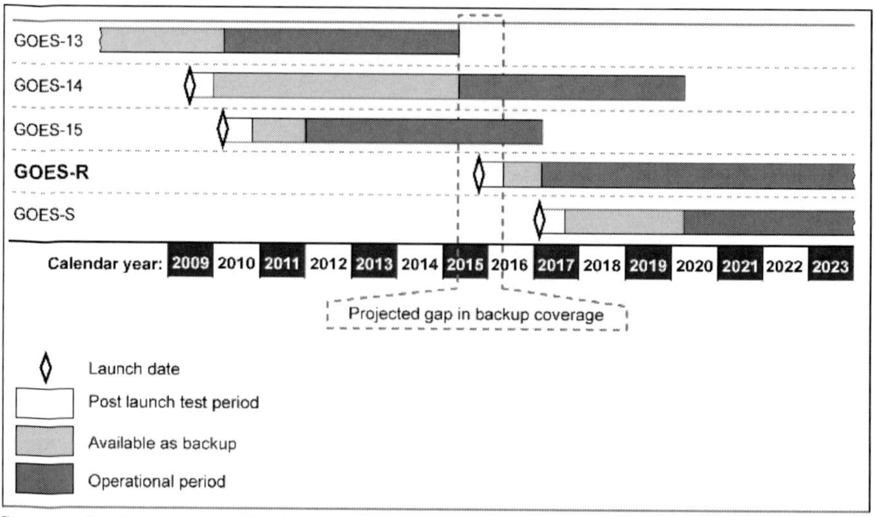

Source: GAO analysis of NOAA data.

Figure 5. Potential Gap in Geostationary Operational Environmental Satellite Coverage.

NOAA Has a Process for Managing Changes in GOES-R Requirements, but Changes Could Affect Some Users

System requirements describe the functionality needed to meet user needs and perform as intended in an operational environment. According to leading industry, academic, and government entities, a disciplined process for developing and managing requirements can help reduce the risks of developing or acquiring a system.[22] One key aspect of effective requirements management involves managing changes to requirements through a standardized process. Table 12 outlines best practices of a sound change management process and key questions for evaluating the process.

The GOES-R program has a change management process that satisfied three practices, partially satisfied two practices, and did not satisfy one practice. Specifically, GOES-R has established a change management process that tracks and documents changes in requirements, documents the rationale for the changes as well as the potential impact of the change on cost and schedule, and ensures that changes are reviewed and approved by a change control board. In addition, the program has evaluated the impact of key changes on selected users and communicated with those users. However, as we first reported in 2010, the program is still weak in evaluating the impact of changes on external users who rely on GOES data products and in effectively communicating changes to those satellite data users.[23] Specifically, table 13 outlines how the GOES-R program performed on each of the best practices for managing changes in requirements, and is followed by a more detailed discussion of key shortfalls.

Table 12. Best Practices in Managing Requirements Changes

Practice	Key questions
Manage changes to requirements throughout the life cycle using a standard process	Does the program (or project) have a requirements management plan? Does the program maintain a current and approved set of requirements? Does the program have an approved set of baseline requirements? Does the program's change management process provide guidance for the identification, review, and management of all requirements changes? Do change management processes apply throughout the program's life cycle?
	Does change management documentation, such as meeting notes or change records, indicate that the organization is following its change management policies and procedures?
Document changes to requirements	Does the organization maintain records for all changes? Are all approved requirements changes documented according to a standard process? Are other work products in consistent alignment with requirements changes?
Document rationale for change and analyze impact	Does the program document rationales for proposed changes? Does the program maintain a history of these rationales? Do they analyze the impact of a proposed change to the project and to users in impact assessments? Do these assessments address impacts to cost, schedule, risk, and project capabilities?
Have an approval body with appropriate representation review and approve all requirements changes	Has the program established an approval body for requirements changes and defined its responsibilities? Do change management policies require appropriate representation on the approval body? Do change management policies require that the approval body review and approve all changes? Does documentation show that the approval body reviewed and approved program requirements changes?

Table 12. (Continued)

Practice	Key questions
Ensure that requirements changes are aligned with user needs	Are requirements analyzed according to a standard process to determine if they continue to meet user needs? Do impact assessments show that the requirements remain in alignment with user needs? Has the program traced the changed requirements back to user needs? Has the program verified and validated that changed requirements align with user needs?
Communicate requirements changes to users	When requirements changes occur, are they communicated to end users? Is change information disseminated as part of a standard process?

Source: GAO analysis of government and industry practices.

Table 13. Assessment of Geostationary Operational Environmental Satellite-R Series Program Practices in Managing Changes in Requirements

Practice	Assessment	Discussion
Manage changes to requirements throughout the life cycle using a standard process	Satisfied	The GOES-R program has a requirements management plan and has established a change management process to apply throughout the project's life cycle. In order to change a high-level requirement, the program must follow a detailed process that begins with proposing a change request, evaluating it, and obtaining approval or rejection of the request. The program also maintains an approved set of high-level baseline requirements and updates them regularly in response to requirements changes.
Document changes to requirements	Satisfied	The GOES-R program documents requirements changes in a public change log associated with its high-level requirements document. More detailed information on the changes is tracked in an internal database. The changes documented in the change log align with those documented in the internal tracking database.
Document rationale for change and analyze impact	Partially satisfied	The GOES-R program documented the rationale for individual requirements changes as well as the cost and schedule impact of selected changes. In addition, the program has assessed the impact of key changes on selected users within NOAA. However, the program has not assessed the cost and schedule impact of all changes, and has not assessed the impact of key

Practice	Assessment	Discussion
		changes on external users who rely on GOES satellite data. Program officials noted that they assessed the cost and schedule impact of changes that were expected to negatively impact the program's cost or schedule, and that they focus their impact assessments on users within NOAA because they are considered the primary users.
Have an approval body with appropriate representation review and approve all changes	Satisfied	The GOES-R program has a configuration change board with representation from key NOAA and NASA officials. The board's responsibilities are formalized in program documentation. Further, the board members review and approve requirements changes.
Ensure that requirements changes are aligned with user needs	Not satisfied	The program's change management process does not require taking steps to ensure that changes in requirements are aligned with user needs. and elevate concerns raised by users.
		Specifically, the process does not require officials to trace applicable changes to user needs or to test or simulate whether the change still meets user needs. Moreover, for seven selected changes we reviewed, the program did not demonstrate the steps it took to test or validate the changes to ensure they were aligned with user needs. Program officials noted that they utilize a user working group to communicate changes to users
Communicate requirements changes to users	Partially satisfied	The GOES-R program generally communicates requirements changes to key users within NOAA and NOAA's National Weather Service through mechanisms such as e-mail correspondence and working groups, while it communicates changes to external users through periodic conferences, such as the GOES users' conferences. However, it does not alert external users who rely on GOES data to perform their missions about specific changes in requirements that will likely affect their operations. These external users include the Federal Aviation Administration, the U.S. Department of Agriculture, and the Department of the Navy. Officials at all three agencies reported that they were not informed about key changes in requirements that could affect their operations. Program officials stated that they work through a variety of working groups to try to communicate changes with those who utilize the satellite data.

Source: GAO analysis of NOAA documentation.

While the program generally communicates requirements changes to key users within NOAA's National Weather Service community, it does not communicate as well with satellite data users external to NOAA. Many such users are dependent on GOES satellite data for their respective missions. Officials responsible for working with satellite weather observations at three agencies were unaware of selected changes in GOES-R requirements. For example, the Federal Aviation Administration uses the satellites' data and images to manage air traffic across the country, and the Navy uses the data for oceanic weather forecasting, as well as tactical ocean analysis of regions of interest. They stated that NOAA had not reached out to them to alert them to these changes or ask if the changes would impact them. Similarly, Forest Service officials were concerned that potential changes in spectrum allocations could affect their ability to obtain data from their own ground-based weather observation systems because they currently rely on GOES-R communication channels to obtain this data.

GOES-R program officials noted that they provide regular briefings to the Office of the Federal Coordinator for Meteorology, an interagency council with membership from fifteen federal departments and agencies involved in meteorological activities (including the Departments of Agriculture, Defense, and Transportation) and that the Air Force represents the Department of Defense community on the GOES-R Series Independent Advisory Committee. However, they acknowledged that they cannot ensure that the information they provide is disseminated within the agencies. Further, GOES officials explained that one reason for the distinction between the internal and external users is that the internal users belong to formal working groups and receive regular updates from the GOES-R program, while the other users generally have more informal or indirect connections with the program. Instead of direct communications such as e-mails, the other users may receive information about GOES-R requirements changes from publically available information or through other meteorological partnerships with NOAA. Without consistent and direct communication, users may be unaware of changes to program requirements that are critical to their respective missions. Because GOES-R users across the country have missions that preserve and protect property and life, it is critical that these organizations are made aware of any changes as soon as they are made, so that they can assess and mitigate any significant impacts.

GOES-R Program Has Undergone Multiple Changes in Requirements; Selected Changes Could Affect User Operations

Since 2007, NOAA has changed multiple system requirements on the GOES-R program. These changes involved strengthening or relaxing specifications on selected products, finalizing a decision not to develop 31 products, and modifying programmatic requirements not tied to any individual product. For example, NOAA strengthened specifications for the geographic coverage, image resolution, and refresh rate on a product depicting total precipitable water, and strengthened accuracy specifications for a product depicting cloud layers and heights. NOAA also relaxed specifications to provide less measurement accuracy on a product depicting hurricane intensity, less geographic coverage on a product depicting sea surface temperatures, less resolution on a product that tracks the motion of clouds through the atmosphere, and less timely updates on a product depicting lightning detection. The GOES-R program also documented NOAA's earlier decision not to develop or provide 31 products that it labeled as optional, noting that the products will only be developed if funding becomes available. In addition, programmatic changes include the elimination of 97 percent mission availability as a measure of minimum performance success and the decision not to transmit raw satellite data to users. Table 14 provides an overview of key changes in product and program requirements since 2007.

Table 14. Summary of Key Changes in Product and Program Requirements between 2007 and 2012

Type of Change	November 2007	October 2012
Product	34 products, each with specifications for accuracy, geographic coverage, resolution, and timeliness	34 products, of which: 20 (59%) were modified • 14 had changes in accuracy measurement • 7 had changes in geographic coverage • 3 had changes in horizontal resolution • 8 had changes in refresh rate/latency
	34 optional products	2 optional products were eliminated 1 optional product was combined with another optional product 31 optional products are not being developed

Table 14. (Continued)

Type of Change	November 2007	October 2012
Program	The satellites shall be capable of being configured to accommodate additional instrumentation with minimal redesign of the spacecraft.	Requirement removed
	The GOES-R series is required to meet or exceed the level of capability of the prior series of satellites (GOES-N,O,P) for system continuity.	Requirement removed
	The GOES-R satellites are required to acquire and transmit the raw environmental data to ground stations to allow for the timely and accurate processing of data.	While the program is still required to relay GOES-R sensor data, the requirement to acquire and transmit raw data has been removed.
	GOES-R is required to meet or exceed the prior series of satellites' capabilities for storage of environmental data.	The program is required to make products available to NOAA Archival Data Centers, but capabilities for storing the data are not specified.
	The GOES-R system need date is specified as December 2014.	Requirement removed
	GOES-R is required to achieve "full operational capability," which is defined, in part, as full coverage of the east and west positions.	The requirement for full operational capability was strengthened to include the production and availability of the full product set of satellite data to users.
	Minimum performance success is defined as 97 percent mission availability for collecting, generating, and distributing key products over a defined central coverage zone.	Minimum performance success is redefined as the successful generation and availability of key functions to users. The availability percentage has been removed.
	The operational lifetime of the GOES-R series shall extend through 2028.	The individual GOES-R satellites' lifetimes shall be 5 years in on-orbit storage plus 10 years in operation.
	Requirements for a remote backup facility not specified	Addition of requirements for a remote backup facility
	Failover time to backup satellite or backup ground facility not mentioned	This information is now included.
	Requirements do not specify the locations of the satellites in on-orbit storage	Added requirements that specify the satellites' checkout location and the location of on-orbit satellite storage

Source: GAO analysis of NOAA documentation.

While NOAA officials stated that they believe that only one of the changes that were made since 2007 was significant,[24] internal and external users noted that they found many of the changes to be significant. In addition, selected satellite data users noted concern at the loss of 17 of the optional products that are no longer being developed. The changes that users found significant, along with user reasoning for why these changes are significant, are listed in table 15. GOES-R program officials acknowledged that the National Weather Service and other users will have impacts from the loss or degradation of products, but that it is not always accurate to assume that GOES-R could have met the original requirements. In 2011, an algorithm development executive board reported that several original requirements could not have been met because, among other reasons, they relied on a hyperspectral instrument that was removed from the program, the requirements were poorly stated and it only became evident later that GOES could not support them, and there were scientific limitations on the development of the products that only became evident after development had started. Program officials stated that they have identified alternative methods for obtaining certain products (some outside the scope of GOES-R) and that they are proactively trying to develop alternative products in coordination with users and other development organizations.

Table 15. User Concerns about Key Changes or Deviations in Requirements

Product	Change	User concerns
Cloud top height	Relaxation of accuracy requirements	Navy officials reported that this change will likely cause significant errors, which will reduce the utility of the cloud top height measurements.
Downward shortwave radiation	Relaxation of accuracy requirements	Navy officials reported that the larger accuracy ranges might make this product difficult to use in a statistically significant way.
Reflected shortwave radiation	Relaxation of accuracy requirements	Navy officials reported that the larger accuracy ranges might make this product difficult to use in a statistically significant way.
Derived stability indices	Relaxation of resolution requirements	Officials from both the Navy and the Federal Aviation Administration expressed concern about this change.

Table 15. (Continued)

Product	Change	User concerns
		The Federal Aviation Administration reported that the reduction in horizontal resolution might result in reduced forecast accuracy and a reduced ability to verify convection, which is useful for predicting severe storms.
Lightning detection	Reduction in product timeliness	Officials from the Federal Aviation Administration expressed concern about this change. They reported that a delay in refresh times could be significant for aviation operations, especially over water areas that rely on satellite data for coverage. In these areas, lightning will be used as an indicator of storm formation and delays in detection and transmission could impact situational awareness.
Magnetometer (geomagnetic field)	Reduction of magnetic field accuracy requirements	The National Weather Service's Space Weather Prediction Center found this change acceptable for the purposes of GOES-R data, but determined that the reduction of the accuracy requirements would noticeably increase error in the instrument's readings of solar energy and the geomagnetic field.
Aerosol particle size	An optional product; not planned to be developed or provided.	Officials from the Department of Agriculture's Forest Service and the Navy expressed concern about not receiving this product. The Forest Service reported that this product would help them monitor and manage air quality.
Aircraft icing threat	An optional product; not planned to be developed or provided.	Officials from the National Weather Service's Aviation Weather Center, the Navy, and the Federal Aviation Administration expressed concern about not receiving this product. The Aviation Weather Center reported that this product would be useful because icing is a major hazard for safe air travel.
Cloud layers / heights	An optional product; not planned to be developed or provided.	Officials from the National Weather Service's Aviation Weather Center and the Navy expressed concern about not receiving this product. The Aviation Weather Center reported that this product would help prevent aviation accidents caused by low visibility.

Product	Change	User concerns
		Low cloud ceiling and visibility was associated with about 20 percent of all aviation accidents from 1994 to 2003.
Cloud liquid water	An optional product; not planned to be developed or provided.	Officials from the National Weather Service's Aviation Weather Center and the Navy expressed concern about not receiving this product. The Aviation Weather Center reported that this product might help identify regions with low visibility. Because low visibility is associated with airline accidents, this product would help prevent aviation accidents.
Cloud type	An optional product; not planned to be developed or provided.	Officials from the National Weather Service's Aviation Weather Center, the Navy, and the Federal Aviation
		Administration expressed concern about not receiving this product. The Aviation Weather Center reported that, as it is related to the icing threat, this product would help air traffic controllers know if a cloud was made of ice, water, or a mixture of the two.
Convective initiation	An optional product; not planned to be developed or provided.	Officials from the National Weather Service's Aviation Weather Center expressed concern about not receiving this product because new convection is critical to air traffic flow management, and convection is a major hazard for safe and efficient flight. In addition, officials from the Navy, the Federal Aviation Administration, and the Department of Agriculture's Forest Service were concerned by the loss of this product. The Forest Service officials reported that the loss of this product could impact its ability to locate potential ignition areas for wildland fires. The National Weather Service's Storm Prediction Center also stated that this product was likely to have had a positive impact on its mission, which is to predict and monitor high impact weather events such as tornadoes.
Enhanced "V"/ overshooting top detection	An optional product; not planned to be developed or provided.	Officials from the National Weather Service's Aviation Weather Center, the Navy, and the Federal Aviation Administration expressed concern about not receiving this product.

Table 15. (Continued)

Product	Change	User concerns
		The Aviation Weather Center reported that this product would indicate the location of turbulence and convection, thereby helping to improve the safety and efficiency of air travel.
		The National Weather Service's Storm Prediction Center also stated that this product was likely to have had a positive impact on its mission.
Flood/standing water	An optional product; not planned to be developed or provided.	Officials from the Department of Agriculture's Forest Service and the Navy expressed concern about not receiving this product. Forest Service officials are concerned that the removal of this product would impact their management of and response to hazards and disasters.
Ice cover	An optional product; not planned to be developed or provided.	Officials from the National Weather Service's Environmental Modeling Center and the Navy expressed concern about not receiving this product. The Environmental Modeling Center reported that ice cover data would help assimilate data received from the sounding sensors.
Low cloud and fog	An optional product; not planned to be developed or provided.	Officials from the National Weather Service's Aviation Weather Center, the Navy, and the Federal Aviation Administration expressed concern about not receiving this product. The Aviation Weather Center reported that this product would help prevent aviation accidents caused by low visibility. Low ceiling and visibility accounted for about 20 percent of all aviation accidents from 1994 to 2003.
Ozone total	An optional product; not planned to be developed or provided.	Officials from the Department of Agriculture's Forest Service are concerned that the removal of this product would impact its ability to monitor and manage air quality.
Probability of rainfall	An optional product; not planned to be developed or provided.	Officials from the National Weather Service's Aviation Weather Center, the Navy, and the Federal Aviation Administration expressed concern about not receiving this product.

Product	Change	User concerns
		The Aviation Weather Center reported that the loss of this product is significant because heavy rainfall relates to air traffic planning and the efficiency of airport operations, and heavy rainfall is correlated with low ceiling and low visibility and/or convection.
		Officials from the Department of Agriculture were also concerned that the loss of this product would impact their predictive services.

Source: GAO analysis of federal agency responses.

In addition to the changes that have already been implemented on the GOES-R program, there are other potential changes that could occur. For example, by the end of 2013, the program plans to decide whether or not to include the Geostationary Lightning Mapper on the GOES-R satellite. Also, there could be changes in the spectrum allocated to weather satellite data. Officials from the National Weather Service and Forest Service raised concerns that these potential changes could also affect their operations. Because these changes have the potential to impact satellite data user operations, it is critical that the GOES-R program communicates program changes to the extended user community. By doing so, satellite data users can establish plans to mitigate any shortfalls in data and minimize the impact of the changes on their operations.

NOAA Developed GOES-R Contingency Plans, but Weaknesses Increase the Impact of a Potential Coverage Gap

GOES satellite data are considered a mission-essential function because of their criticality to weather observations and forecasts. These forecasts—such as those for severe storms, hurricanes, and tornadoes— can have a substantial impact on our nation's people, infrastructure, and economy. Consequently, NOAA policy requires that there must be two in-orbit GOES satellites and one on-orbit spare in operation at all times. If one of the operational satellites were to fail, the on-orbit spare could be moved into position to take the place of the failed satellite. However, if there are delays in the launch of the GOES-R satellite or if either of the two satellites currently in operation were to fail, NOAA would not have an on-orbit spare to fill the gap.

Table 16. Guidelines for Developing a Sound Contingency Plan

Category	Key elements
Identifying failure scenarios and impacts	Define likely failure scenarios Conduct impact analyses showing impact of failure scenarios on business processes and user requirements Define minimum acceptable level of outputs and recovery time objectives, and establish resumption priorities
Developing contingency plans	Define roles and responsibilities for implementing contingency plans Identify alternative solutions to address failure scenarios Select contingency strategies from among alternatives based on costs, benefits, and impacts Develop "zero-day" procedures Define actions needed to implement contingency strategies Define and document triggers and time lines for enacting the actions needed to implement contingency plans Ensure that steps reflect priorities for resumption of products and recovery objectives Designated officials review and approve contingency plan
Validating and implementing contingency plans	Identify steps for testing contingency plans and conducting training exercises Prepare for and execute tests Execute applicable actions for implementation of contingency strategies Validate test results for consistency against minimum performance levels Communicate and coordinate with stakeholders to ensure that contingency strategies remain optimal for reducing potential impacts Update and maintain contingency plans as warranted

Source: GAO analysis of guidance documents from the National Institute of Standards and Technology, Software Engineering Institute, GAO, NOAA, and the GOES-R program.

Government and industry best practices call for the development of contingency plans to maintain an organization's essential functions in the case of an adverse event.[25] These practices include key elements such as defining failure scenarios, identifying and selecting strategies to address failure scenarios, developing procedures to implement the selected strategies, identifying any actions needed to implement the strategies, testing the plans, and involving affected stakeholders. These elements can be grouped into categories, including (1) identifying failure scenarios and impacts, (2) developing contingency plans, and (3) validating and implementing contingency plans (see table 16).

NOAA has established contingency plans for both its GOES satellites and its associated ground systems. In September 2010, we recommended that NOAA develop and document continuity plans for the operation of geostationary satellites that include the implementation procedures, resources, staff roles, and timetables needed to transition to a single satellite, a foreign satellite, or other solution. In September 2011, the GOES-R program provided a draft plan documenting a strategy for conducting operations if there were only a single operational satellite. In December 2012, the program provided us with a final version of this plan. It included scenarios for three, two, and one operational satellites. In addition to this satellite contingency plan, NOAA has another contingency-related plan with activation procedures for its satellites.

Furthermore, the NOAA office responsible for ground-based satellite operations and products has created plans for contingency operations at the GOES ground system facility, the Satellite Operations Control Center.

Specifically, NOAA's plans describe the transfer of critical functions to a backup facility during an emergency. The continuity plan contains, among other things, descriptions of the alternate locations for resources, and the performance of key functions and implementation procedures.

When compared to best practices, NOAA's satellite and ground system contingency plans had many strengths and a few weaknesses. Specifically, the satellite contingency plan fully implemented seven elements, partially implemented nine elements, and did not implement one element. The ground system contingency plan fully implemented ten elements, partially implemented six elements, and one element was not applicable. Table 17 shows the extent to which the satellite and ground system contingency plans fully implemented, partially implemented, or did not implement key contingency planning elements.

NOAA has implemented most of the best practices on both the GOES satellite and ground contingency plans. Specifically, NOAA identified failure scenarios, recovery priorities, and minimum levels of acceptable performance. NOAA also established contingency plans that identify solutions and high-level activities and triggers to implement the solutions. Further, the agency has tested its contingency plans, trained staff on how to implement the contingency plans, and updated the plans when warranted. The agency also successfully implemented its contingency plans when it experienced problems with one of its operational satellites.

Table 17. Implementation of Key Contingency Planning Elements for Geostationary Operational Environmental Satellites

Category	Key element	Satellite system	Ground system	Description
Identifying failure scenarios and impacts	Define likely failure scenarios	Fully implemented	Fully implemented	NOAA has defined three likely failure scenarios for its satellite system—the loss of one, two, or all three satellites in the GOES constellation. The agency also defines the conditions that would constitute a satellite failure. NOAA's scenarios are broad enough that they cover a wide range of situations, including a gap caused by a delay in the GOES-R launch. NOAA has defined likely ground system failure scenarios.
	Conduct impact analyses showing impact of failure scenarios on business processes and user requirements	Not implemented	Partially implemented	NOAA did not conduct impact analyses showing the impact of satellite failure scenarios on business processes or user requirements. NOAA conducted impact analyses of ground system outages and disruptions on business processes and user requirements; however, these analyses do not reflect each failure scenario.
	Define minimum acceptable level of outputs and recovery time objectives, and establish resumption priorities	Fully implemented	Fully implemented	NOAA defined minimum acceptable output criteria for satellites, instruments and products in its satellite plans as well as for business processes and subsystems in the ground system plans.
Developing contingency plans	Define roles and responsibilities for implementing contingency plans	Partially implemented	Partially implemented	NOAA has defined roles and responsibilities for some, but not all, contingency operations in both the satellite and ground system plans. For example, the satellite contingency plan identifies roles and responsibilities for briefing management in the event of losing an operational satellite, but does not define responsibility for notifying users.

Category	Key element	Satellite system	Ground system	Description
				The ground system contingency plans describe roles and responsibilities of three contingency teams, but do not clearly define the roles and responsibilities for the contingency coordinator.
	Identify alternative solutions to address failure scenarios	Partially implemented	Fully implemented	In its satellite contingency plan, NOAA identified alternative solutions to address satellite failure scenarios, including relocating and using older GOES satellites and requesting coverage by foreign satellites. However, NOAA did not identify alternative solutions for preventing delays in the GOES-R launch, which could cause a reduction in the number of satellites. For its ground systems, NOAA identified a solution for its failure scenarios: to switch operations to one of several backup locations.
	Select contingency strategies from among alternatives based on costs, benefits, and impacts	Partially implemented	Partially implemented	In both sets of plans, NOAA has selected contingency strategies to address failure scenarios; however, it did not provide evidence that it had selected these strategies from alternatives based on costs, benefits, and impacts. Moreover, NOAA did not select strategies to prevent one of the most likely situations that would trigger a failure scenario: a delay in the launch of the GOES-R satellite.
	Develop "zero-day" procedures	Partially implemented	Fully implemented	NOAA identified strategies and procedures for addressing GOES satellite failure scenarios, but did not establish associated time frames. NOAA developed zero-day strategies and procedures for the GOES ground system.
	Define actions needed to implement contingency strategies	Partially implemented	Fully implemented	NOAA has defined high-level activities to implement satellite contingency strategies, such as relocation of a satellite to a central location and user notification of a switch to a single satellite—however, no detailed procedure steps are given for performance of these activities. NOAA has defined the steps to implement GOES ground system contingency strategies.

Table 17. (Continued)

Category	Key element	Satellite system	Ground system	Description
	Define and document triggers and time lines for enacting the actions needed to implement contingency plans	Partially implemented	Partially implemented	NOAA has identified triggers and specific time lines for implementing satellite contingency plans. However, it has not established triggers or time lines for any actions it might take to prevent a delay in the GOES-R launch. NOAA has identified two different triggers for enacting the ground system plan, but the plan does not describe which trigger is to be used.
	Ensure that steps reflect priorities for resumption of products and recovery objectives	Partially implemented	Partially implemented	NOAA's satellite contingency plan describes its recovery objectives and prioritizes GOES instruments and products; however, the steps for implementing contingency strategies do not reflect these priorities and objectives. Ground system contingency strategies establish priorities for resuming operations, but do not define recovery time objectives.
	Designated officials review and approve contingency plan	Fully implemented	Fully implemented	A designated official has reviewed and approved both sets of contingency plans.
Validating and implementing contingency plans	Identify steps for testing contingency plans and conducting training exercises	Fully implemented	Fully implemented	NOAA has identified steps for testing GOES satellite contingency plans and has conducted exercises and simulations. NOAA has also identified steps for testing and conducting exercises and simulations on its ground system contingency plans. NOAA provides training to its operations staff on contingency operations for both the satellite and ground systems.
	Prepare for and execute tests	Fully implemented	Fully implemented	NOAA officials provided documentation showing preparation for and execution of regular maneuvers of on-orbit satellites. According to officials, these maneuvers are similar to the maneuvers identified as an action in the contingency plans.

Category	Key element	Satellite system	Ground system	Description
				NOAA also prepared for and executed tests of its ground system contingency plans.
	Execute applicable actions for implementation of contingency strategies	Fully implemented	Not Applicable	NOAA has performed actions to implement contingency strategies, including activities to monitor the health and safety of the satellites, and to provide status information to management. Executing actions is not applicable for the ground system contingency plan, because that plan does not identify actions to be taken at the present time.
	Validate test results for consistency against minimum performance levels	Partially implemented	Fully implemented	NOAA tested a series of satellite maneuvers similar to those that would be used in the event of a failure, but did not demonstrate how these or other scenario tests would meet minimum performance levels. On the ground system, NOAA performed tests to validate contingency operations, and demonstrated that the transfer of responsibility meets minimum recovery performance levels.
	Communicate and coordinate with stakeholders to ensure that contingency strategies remain optimal for reducing potential impacts	Partially implemented	Partially implemented	According to users, NOAA is proactive in communicating potential changes and impacts when issues develop, and responded quickly to a recent outage in a GOES satellite. However, the contingency strategies currently in place of (1) switching to single satellite operations and (2) using a foreign satellite as a temporary replacement would have a major effect on user operations; NOAA has not provided key external users with information on meeting data needs under these scenarios. For example, the Forest Service relies on GOES satellites to obtain data from its distributed ground-based observation network, but NOAA has not discussed potential mitigation options specific to this scenario.
	Update and maintain contingency plans as warranted	Fully implemented	Fully implemented	NOAA has updated and maintained contingency plans for both the GOES satellites and ground system.

Source: GAO analysis of NOAA documentation.

Specifically, when GOES-13 experienced problems in September and October 2012, NOAA activated its contingency plans to move its back-up satellite into position to provide observations until GOES-13 was once again operational.

While the agency has not needed to address the loss of a back-up satellite in recent years, contingency plans cover this situation by determining if older GOES satellites could provide coverage, moving the single satellite into a central position over the country, and seeking data from foreign satellites.

However, both satellite and ground contingency plans contain areas that fall short of best practices. For example, NOAA has not demonstrated that the contingency strategies for both its satellite and ground system are based on an assessment of costs, benefits, and impact on users. Further, the satellite plan does not specify procedures for working with the user community to account for potential reductions in capability under contingency operations. For example, officials from the Federal Aviation Administration noted that NOAA's contingency plans do not define the compatibility, security, and standard protocol language they should use if a foreign satellite were to be utilized. Also, while selected users reported that, in the past, they have been well informed by NOAA when changes in service occur, including the problems with GOES-13, others were either not informed or received information on outages through a third party. Moreover, selected users stated that certain contingency operations could have a significant impact on their operations. For example, Federal Aviation Administration officials stated that flight approaches in Alaska that were enabled using the Global Positioning System were affected by the GOES-13 outage in late 2012. As another example, Forest Service officials explained that if GOES were to experience an outage and not have a backup satellite available, it was their understanding that NOAA would either move a single satellite into a central position over the country or obtain observations from a foreign satellite. Under both of these scenarios, they could lose views of wildland fires and their ability to obtain data from ground-based observation networks. Nearly all users stated that the effects of a switch to a single satellite or foreign satellite configuration would be significant.

In addition, while NOAA's failure scenarios for its satellite system are based on the number of available satellites—and the loss of a backup satellite caused by a delayed GOES-R launch would fit into these scenarios—the agency did not identify alternative solutions or time lines for preventing a GOES-R launch delay. According to NOAA officials, a gap caused by a delayed launch would trigger the same contingency actions as a failure on launch or the loss of a currently on-orbit satellite. However, this does not take into account potential actions that NOAA could undertake to prevent a delayed launch, such as removing selected functionality or compressing test schedules.

NOAA officials stated that their focus on primary users and on the number of available satellites is appropriate for their contingency plans. Given the potential for a delay in the launch of the GOES-R satellite and the expectation that there will be at least a year with no backup satellite in orbit, it is important that NOAA consider ways to prevent a delay in the GOES-R launch, and ensure its contingency plans are fully documented, tested, and communicated to affected stakeholders. Further, it is critical that NOAA and users are aware of how contingency scenarios will affect user operations. Until comprehensive plans are developed, it is less certain that NOAA can provide a consistent level of service and capabilities in the event of an early failure or late launch. This in turn could have a devastating effect on the ability of meteorologists to observe and report on severe weather conditions.

CONCLUSION

The GOES-R program is well on its way toward developing the first satellite in the series, but it continues to face risks that could delay the first satellite's launch. Among these risks are issues we have previously raised on how the program manages reserve funds and implements sound scheduling practices. Specifically, the agency does not provide important details on its contingency reserve funds to senior executives, including the reserves allocated for each of the four satellites or key assumptions made in calculating reserves. Without this information, program officials could misinterpret the size of the remaining reserves and make poor decisions regarding the program's future funding. The agency has improved selected scheduling practices, but others remain weak—in part, according to agency officials, due to the dynamic nature of scheduling a program as complex as the GOES-R satellite program. As the agency closes in on its expected launch date, technical issues in developing the space and ground segments and scheduling problems could make it more difficult to launch on schedule, and program officials now acknowledge that the launch date may be delayed by 6 months. Any delay in the anticipated launch date would expand a potential one-year gap in the availability of an on-orbit backup GOES satellite, and raise the risk of a gap in geostationary satellite data should one of the two operational satellites experience a problem.

While the agency has made multiple changes to GOES-R requirements in recent years, it has not effectively involved satellite data users in those changes. Specifically, internal NOAA and external satellite data users were not fully informed about changes in GOES-R requirements and did not have a chance to communicate their concerns about the impact these changes could have on their ability to perform their missions. Many of these users expressed concerns about the effect these changes could have on their ability to fulfill their missions, including facilitating air traffic, conducting military operations, and fighting wildland fires. Until NOAA improves its outreach and communication with external satellite data users, its changes in requirements could cause unexpected impacts on critical user operations.

Given the possibility of a gap in geostationary satellite coverage, NOAA has established contingency plans for both its GOES satellites and ground systems; these plans include the likely scenario in which there will not be an on-orbit backup. While these plans include many elements called for in industry best practices, the satellite contingency plan did not assess the potential impacts of a failure on users, or specify actions for working with the user community to address these potential reductions in capability under contingency operations. They also did not identify alternative solutions or time lines for preventing a delay in the GOES-R launch date. The absence of a fully-tested and complete set of GOES-Rrelated contingency plans and procedures could have a major impact on levels of service provided in the event of a satellite or ground system failure.

RECOMMENDATIONS FOR EXECUTIVE ACTION

To address risks in the GOES-R program development and to help ensure that the satellite is launched on time, we are making the following four recommendations to the Secretary of Commerce. Specifically, we recommend that the Secretary of Commerce direct the NOAA Administrator to:

- Direct program officials to include information on the amount of reserve funding for each of the four satellites in the program as well as information on the calculation and use of reserves in regular briefings to NOAA senior executives, so that executives are fully informed about changes in reserve levels.
- Given the likely gap in availability of an on-orbit GOES backup satellite in 2015 and 2016, address the weaknesses identified in this

report on the core ground system and the spacecraft schedules. These weaknesses include, but are not limited to, sequencing all activities, ensuring there are adequate resources for the activities, and conducting a schedule risk analysis.

- Improve communications with internal and external satellite data users on changes in GOES-R requirements by (a) assessing the impact of changes on user's critical operations; (b) seeking information from users on any concerns they might have about past or potential changes; and (c) disseminating information on past and potential changes in requirements to satellite data users.
- Revise the satellite and ground system contingency plans to address weaknesses identified in this report, including providing more information on the potential impact of a satellite failure, identifying alternative solutions for preventing a delay in GOES-R launch as well as time lines for implementing those solutions, and coordinating with key external stakeholders on contingency strategies.

AGENCY COMMENTS AND OUR EVALUATION

We sought comments on a draft of our report from the Department of Commerce and NASA. We received written comments on a draft of this report from Commerce transmitting NOAA's comments. NOAA concurred with all four of our recommendations and identified steps that it is taking to implement them. It also provided technical comments, which we have incorporated into our report, as appropriate.

While NOAA concurred with our recommendation to include information on reserve funding for each of the four satellites in the program and information on the calculation and use of reserves in regular briefings to senior executives, and suggested that its current processes fulfill this recommendation, we do not believe they do. Specifically, NOAA stated that the GOES-R program currently reports on reserve funding at two major monthly management meetings, which alerts management if reserves fall below designated thresholds for the remaining work on all four satellites. The agency also stated that its reporting of the percent of "unliened" contingency funding—the amount of contingency funding not allocated to a potential risk or issue—for the remaining work addresses our concern regarding whether there are sufficient reserves to complete the GOES-R series.

However, the GOES-R program does not currently identify the reserve funding needed for each individual satellite or provide details on how reserves are being calculated and used at the monthly management meetings. By not providing reserve information on the individual satellites, the program is not alerting management about potential near-term funding shortfalls. For example, maintaining a high level of reserves on the later satellites could mask a low level of reserves in the near-term for GOES-R and S. Such a scenario could affect the satellites' development schedules and launch dates. Further, by not obtaining details on the assumptions made when calculating reserves and the causes of changes in reserve values, management is unable to determine if changes in reserve levels are due to the addition, subtraction, or use of funds, or to changes in the assumptions used in the calculations. Given the importance of reserve funds in ensuring the satellite development remains on track, management should be aware of reserve funding levels for each individual satellite and of the underlying reasons for changes in reserve levels. Therefore, we continue to believe that additional action is needed by NOAA to respond to our recommendation.

After we received agency comments and while our report was in final processing, NOAA notified us that the launch dates of the first and second GOES-R series satellites would be delayed. Given the late receipt of this information, our report reflects the previous launch date.

NASA did not provide comments on the report's findings or recommendations, but noted that it would provide any input it might have to NOAA for inclusion in that agency's comments.

David A. Powner
Director,
Information Technology Management Issues

Appendix I: Objectives, Scope, and Methodology

Our objectives were to (1) assess the National Oceanic and Atmospheric Administration's (NOAA) progress in developing the Geostationary Operational Environmental Satellite-R series (GOES-R) program and in addressing key cost and schedule risks that we identified in a prior report, (2) evaluate the program's efforts to manage changes in requirements and whether any significant changes have recently occurred, and (3) evaluate the adequacy of GOES-R contingency plans.

To assess NOAA's progress in developing the GOES-R satellite program, we compared the program's planned completion dates for key milestones identified in its management control plan and system review plan against actual and currently estimated completion dates. We analyzed monthly program status briefings to identify the current status and recent development challenges of flight and ground project components and instruments. To assess NOAA's efforts to address key cost risks, we compared program-reported data on development costs and reserves to best practices in reserve funding as identified by the program's management control plan, which, in turn, reflects National Aeronautics and Space Administration requirements. We calculated reserve percentages using program office data on development costs and reserves, and compared these calculations to the reserve percentages reported by the program to management. To assess NOAA's efforts to address key schedule risks, we compared schedules for two key GOES-R components to best practices in schedule development as identified in our Cost Estimating and Assessment Guide.[1] Similar to our previous report, we used a five-part rating system. We then compared our previous assessment to our current assessment to identify practices that were improved, stayed the same, or became weaker over time. We conducted interviews with GOES-R program staff to better understand milestone time frames, to discuss current status and recent development challenges for work currently being performed on GOES-R, and to understand how the program reports costs and reserve totals. We also examined the reliability of data on cost reserves and program schedules. Regarding cost reserves, we examined reliability by recalculating reserve percentages based on supporting data over a period of one year, and compared the results to those presented by the program to management. Regarding schedules, we created a template that examined each schedule in areas such as missing logic, tasks completed out of sequence, and completed tasks with start or finish dates in the future. As a result, we found both the reserve information and the schedules to be reliable for the purposes of conducting our analyses.

To evaluate the program's efforts to manage changes in requirements, we compared GOES-R practices for managing requirements changes against best practices, which we drew from several leading industry sources including the Software Engineering Institute's Capability Maturity Model®–Integration, the Program Management Institute's Program Manager's Body of Knowledge, the Federal Information Security Controls Audit Manual and the Information Technology Governance Institute's Control Objectives for Information and related Technology governance framework. We assessed GOES-R practices as having satisfied, partially satisfied, or not satisfied each best practice. We analyzed changes from 2007 to the present in the program's Level I Requirements Document to determine the extent of the changes. We also identified concerns about these changes from a subset of satellite data users. We selected users from both inside and outside NOAA's National Weather Service, the main GOES satellite user, based on several factors: the importance of GOES data to the organization's core mission, the user's reliance on GOES products that have changed or may change, and—for agencies outside of NOAA—the percentage of spending devoted to meteorological operations. The user organizations outside of NOAA included in our review were: the US Department of Agriculture, the Department of Transportation's Federal Aviation Administration, and the Department of Defense's Navy and Air Force. User organizations inside of NOAA's National Weather Service included the Aviation Weather Center, Space Weather Prediction Center, Storm Prediction Center, Environmental Modeling Center, and a Weather Forecast Office.

To evaluate the adequacy of GOES-R contingency plans, we compared contingency plans and procedures for both GOES satellites and the GOES ground system against best practices developed from leading industry sources such as the National Institute of Standards and Technology, the Software Engineering Institute's Capability Maturity Model®–Integration, and our prior work. We analyzed the contingency plans to identify strategies for various failure scenarios and determined whether the satellite and ground system contingency plans fully implemented, partially implemented, or did not implement each of the practices. We also interviewed selected satellite data users to better determine the impact of a GOES failure scenario on their operations, and the level of communication they have had with NOAA satellite offices on current contingency plans.

We performed our work at NOAA, National Aeronautics and Space Administration, and US Department of Agriculture offices in the Washington, D.C., area and at National Weather Service offices in Kansas City, Missouri; Norman, Oklahoma; and Sterling, Virginia. We conducted this performance audit from October 2012 to September 2013, in accordance with generally accepted government auditing standards. Those standards require that we plan and perform the audit to obtain sufficient, appropriate evidence to provide a reasonable basis for our findings and conclusions based on our audit objectives. We believe that the evidence obtained provides a reasonable basis for our findings and conclusions based on our audit objectives.

End Notes

[1] GAO, GAO Cost Estimating and Assessment Guide: Best Practices for Developing and Managing Capital Program Costs, GAO-09-3SP (Washington, D.C.: Mar. 2009); NOAA, Geostationary Operational Environmental Satellites—R Series Management Control Plan (Silver Spring, Md.: January 2013).

[2] NASA, NASA Systems Engineering Handbook (Washington, D.C.: December 2007), Software Engineering Institute, CMMI® for Development, Version 1.3 (Pittsburgh, Pa.: November 2010); Project Management Institute, A Guide to the Project Management Body of Knowledge (Newtown Square, Pa.: 2004); GAO, Federal Information System Controls Audit Manual, GAO-09-232G (Washington, D.C.: February 2009); IT Governance Institute, Control Objectives for Information and related Technology 4.1 (Rolling Meadows, Ill.: 2007).

[3] GAO, Year 2000 Computing Crisis: Business Continuity and Contingency Planning, GAO/AIMD-10.1.19 (Washington, D.C.: August 1998); National Institute of Standards and Technology, Contingency Planning Guide for Federal Information Systems, NIST 800-34 (Gaithersburg, Md.: May 2010); Software Engineering Institute, CMMI® for Acquisition, Version 1.3 (Pittsburgh, Pa.: November 2010).

[4] While our report was in final processing, NOAA announced that it would delay the launch dates for its GOES-R and GOES-S satellites to the second quarter of fiscal year 2016 and the third quarter of fiscal year 2017, respectively.

[5] GAO, Environmental Satellites: Focused Attention Needed to Mitigate Program Risks, GAO-12-841T, (Washington, D.C.: June 27, 2012); GAO, Geostationary Weather Satellites: Design Progress Made, but Schedule Uncertainty Needs to be Addressed, GAO-12-576, (Washington, D.C.: June 26, 2012); GAO, Geostationary Operational Environmental Satellites: Improvements Needed in Continuity Planning and Involvement of Key Users, GAO-10-799 (Washington, D.C.: Sept. 1, 2010); GAO, Geostationary Operational Environmental Satellites: Acquisition Has Increased Costs, Reduced Capabilities, and Delayed Schedules, GAO-09-596T (Washington, D.C.: Apr. 23, 2009); GAO, Geostationary Operational Environmental Satellites: Acquisition Is Under Way, but Improvements Needed in Management and Oversight, GAO-09-323 (Washington, D.C.: Apr. 2, 2009); GAO, Geostationary Operational Environmental Satellites: Further Actions Needed to Effectively Manage Risks, GAO-08-183T (Washington, D.C.: Oct. 23, 2007); GAO, Geostationary Operational Environmental Satellites: Progress Has Been Made, but Improvements Are Needed to Effectively Manage Risks, GAO-08-18 (Washington, D.C.: Oct. 23, 2007); GAO, Geostationary Operational Environmental Satellites: Additional Action Needed to Incorporate Lessons Learned from Other Satellite Programs, GAO-06-1129T (Washington, D.C.: Sept. 29, 2006); and GAO, Geostationary Operational Environmental Satellites: Steps Remain in Incorporating Lessons Learned from Other Satellite Programs, GAO-06-993 (Washington, D.C.: Sept. 6, 2006).

[6] GAO-10-799.

[7] GAO-10-799.

[8] GAO, 2013 High-Risk Series: An Update, GAO-13-359T (Washington, D.C.: February 14, 2013).

[9] GAO-12-756.

[10] This module is called the Mission Management function.

[11] GAO, GAO Cost Estimating and Assessment Guide: Best Practices for Developing and Managing Capital Program Costs, GAO-09-3SP (Washington, D.C.: Mar. 2009).

[12] While NOAA has ultimate responsibility for GOES-R, NOAA shares program management responsibilities with NASA, and the program office is located at NASA's Goddard Space Flight Center.

[13] Until late 2012, NOAA required the ground project to maintain 30 percent of its development cost as a reserve. However, program officials recently revised the requirement down to 20 percent to reflect the shorter amount of development time before launch and the retirement of some risks.

[14] GAO-12-576.

[15] Under testing, the electronics board emitted unexpectedly high levels of radiation, which would cause a high number of false alarms and hinder the program's ability to assess the instrument's observations.

[16] While our report was in final processing, NOAA announced that it would delay the launch date for its GOES-R satellite from October 2015 to the second quarter of fiscal year 2016.

[17] See GAO-09-3SP. In May 2012, we published updated guidance on scheduling best practices. See GAO, Schedule Assessment Guide: Best Practices for Project Schedules— Exposure Draft, GAO-12-120G (Washington, D.C.: May 30, 2012). The updated guidance identifies 10 best practices.

[18] GAO-12-576.

[19] Level-of-effort activities represent work that has no measurable output and cannot be associated with a physical product or defined deliverable. These activities are typically related to management and other oversight that continues until the detailed activities they support have been completed.

[20] Total float time is the amount of time an activity can be delayed or extended before the delay affects its successors or the program's finish date.

[21] A driving path is the longest path of successive activities that drives the finish date for a key milestone. The driving path often corresponds to a schedule's critical path.

[22] Leading industry and governments sources—including the Software Engineering Institute's Capability Maturity Model®–Integration, the Project Management Institute's Project Management Body of Knowledge, the Federal Information Security Controls Audit Manual, the IT Governance Institute's Control Objectives for Information and related Technology governance framework, and NASA system development policies—provide extensive guidance on managing requirements.

[23] GAO-10-799.

[24] NOAA officials stated that the sole significant change was a reduction in the accuracy requirement for the magnetometer, and demonstrated that they had obtained approval from the most pertinent user community, the National Weather Service's Space Weather Prediction Center, before making the change.

[25] See GAO, Year 2000 Computing Crisis: Business Continuity and Contingency Planning, GAO/AIMD-10.1.19 (Washington, D.C.: August 1998); National Institute of Standards and Technology, Contingency Planning Guide for Federal Information Systems, NIST 800-34 (Gaithersburg, Md.: May 2010); Software Engineering Institute, CMMI® for Acquisition, Version 1.3 (Pittsburgh, Pa.: November 2010).

End Notes for Appendix I

[1] See GAO, GAO Cost Estimating and Assessment Guide: Best Practices for Developing and Managing Capital Program Costs, GAO-09-3SP (Washington, D.C.: Mar. 2009). In May 2012, GAO published updated guidance on scheduling best practices. See GAO, Schedule Assessment Guide: Best Practices for Project Schedules—Exposure Draft, GAO-12-120G (Washington, D.C.: May 30, 2012). The updated guidance identifies 10 best practices. In order to compare past and current results, we conducted our current assessment using the original 9 practices.

In: Weather Satellites
Editor: Sherril G. Ryder

ISBN: 978-1-62948-687-1

Chapter 2

POLAR WEATHER SATELLITES: NOAA IDENTIFIED WAYS TO MITIGATE DATA GAPS, BUT CONTINGENCY PLANS AND SCHEDULES REQUIRE FURTHER ATTENTION*

United States Government Accountability Office

WHY GAO DID THIS STUDY

NOAA established the JPSS program in 2010 to replace aging polar satellites and provide critical environmental data used in forecasting weather and measuring variations in climate. However, program officials anticipate a gap in satellite data between the time that the S-NPP satellite reaches the end of its life and the JPSS-1 satellite becomes operational (see graphic). Given the criticality of satellite data to weather forecasts, the likelihood of a significant satellite data gap, and the potential impact of a gap on the health and safety of the U.S. population and economy, GAO added this issue to its High Risk List in 2013.

GAO was asked to review the JPSS program because of the importance of polar satellite data. GAO's objectives were to (1) evaluate NOAA's progress in

---

* This is an edited, reformatted and augmented version of United States Government Accountability Office, Publication No. GAO-13-676, dated September 2013.

sustaining the continuity of NOAA's polar-orbiting satellite system through S-NPP and JPSS satellites; (2) evaluate the quality of NOAA's program schedule; and (3) assess NOAA's plans to address potential gaps in polar satellite data. To do so, GAO analyzed program management status reports, milestone reviews, and schedule data; examined polar gap contingency plans; and interviewed agency and contractor officials.

WHAT GAO RECOMMENDS

GAO is recommending NOAA develop a mechanism to track the usage of its satellite products, establish a complete integrated master schedule, address weaknesses in component schedules, and address shortfalls in polar satellite gap contingency plans. NOAA concurred with GAO's recommendations and identified steps it is taking to implement them.

WHAT GAO FOUND

The National Oceanic and Atmospheric Administration (NOAA) has made noteworthy progress on the Joint Polar Satellite System (JPSS) program by delivering data from its first satellite—the Suomi National Polar-orbiting Partnership (S-NPP)—to weather forecasters, completing significant instrument development for the next satellite (called JPSS-1), and reducing the program's life cycle cost estimate from $12.9 billion to $11.3 billion by refocusing on weather products. However, key challenges remain. Specifically, S-NPP has not yet achieved full operational capability because the program is behind schedule in validating the readiness of satellite products. Also, the program does not track whether key users are using its products or if the products meet the users' needs. In addition, issues with the JPSS ground system schedules have delayed the delivery of key system capabilities. Until the program addresses these challenges, it may continue to experience delays in delivering actionable S-NPP data to system users and in meeting JPSS-1 development schedules.

A program's success depends in part on having an integrated master schedule that defines when and how long work will occur and how activities are related to each other; however, the JPSS program office does not yet have a complete integrated master schedule and weaknesses exist in component schedules. Specifically, the program established an integrated master schedule in June 2013 and is reporting a 70 percent confidence level in the JPSS-1 launch date. However, about one-third of the program schedule is missing information needed to establish the sequence in which activities occur. In addition, selected component schedules supporting the JPSS-1 satellite have weaknesses including schedule constraints that have not been justified. Until the program completes its integrated schedule and addresses weaknesses in component schedules, it will lack the information needed to effectively monitor development progress and have less assurance of meeting the planned JPSS-1 launch date.

While NOAA developed a mitigation plan to address a potential 14 to 18 month gap in afternoon polar satellite data in October 2012 and subsequently identified additional alternatives for addressing potential gaps, it has not yet established a comprehensive contingency plan. Specifically, NOAA has not yet revised its mitigation plan to include the new alternatives, and the plan lacks several key elements, such as triggers for when to take key actions and detailed procedures for implementing them. Until NOAA establishes a comprehensive plan, it may not be sufficiently prepared to mitigate anticipated gaps in polar satellite coverage.

ABBREVIATIONS

A-DCS	Advanced Data Collection System
ATMS	Advanced Technology Microwave Sounder
CERES	Cloud and Earth's Radiant Energy System
CrIS	Cross-Track Infrared Sounder
DMSP	Defense Meteorological Satellite Program
DOD	Department of Defense
IDPS	Interface Data Processing Segment
JCL	Joint Cost and Schedule Confidence Level
JPSS	Joint Polar Satellite System
MetOp	Meteorological Operational (satellite)
NASA	National Aeronautics and Space Administration
NESDIS	National Environmental Satellite, Data, and Information Service
NOAA	National Oceanic and Atmospheric Administration
NPOESS	National Polar-orbiting Operational Environmental Satellite System

S-NPP	Suomi National Polar-orbiting Partnership
OMPS	Ozone Mapping and Profiler Suite
POES	Polar-orbiting Operational Environmental Satellites
TSIS	Total and Spectral Solar Irradiance Sensor
VIIRS	Visible/Infrared Imager/Radiometer Suite

September 11, 2013

The Honorable Lamar Smith
Chairman

The Honorable Ralph Hall
Chairman Emeritus

The Honorable Eddie Bernice Johnson
Ranking Member
Committee on Science, Space,
and Technology
House of Representatives

The National Polar-orbiting Operational Environmental Satellite System (NPOESS) program was planned to be a state-of-the-art, environment-monitoring satellite system that would replace two existing polar-orbiting environmental satellite systems. Managed jointly by the Department of Commerce's National Oceanic and Atmospheric Administration (NOAA), the Department of Defense (DOD)/U.S. Air Force, and the National Aeronautics and Space Administration (NASA), the program was considered critical to the nation's ability to maintain the continuity of data required for weather forecasting and global climate monitoring through the year 2026.

However, in the 8 years after the development contract was awarded in 2002, the NPOESS cost estimate had more than doubled—to about $15 billion, launch dates had been delayed by over 5 years, significant functionality had been removed from the program, and the program's tri-agency management structure had proven to be ineffective. Importantly, delays in launching the satellites put the program's mission at risk. To address these challenges, a task force led by the White House's Office of Science and Technology Policy reviewed the management and governance of the NPOESS program. In February 2010, the Director of the Office of Science and Technology Policy announced a decision to disband the NPOESS acquisition and, instead, have NOAA and DOD undertake separate acquisitions, with NOAA responsible for satellites in the afternoon orbit and DOD responsible for satellites in the early morning orbit. After that decision, NOAA began developing plans for the Joint Polar Satellite System (JPSS). In October 2011, the JPSS program successfully launched the Suomi National Polar-orbiting Partnership (S-NPP) demonstration satellite, the first in a series of satellites to be launched as part of NOAA's JPSS program.

Given your interest in the progress NOAA has made on the JPSS program, our objectives were to (1) evaluate NOAA's progress in meeting program objectives of sustaining the continuity of the polar-orbiting satellite system through the S-NPP and JPSS satellites, (2) evaluate the quality of the JPSS program schedule, and (3) assess NOAA's plans to address potential gaps in polar satellite data.

To evaluate NOAA's progress in meeting its program objectives, we analyzed plans and reports on the satellites' system development efforts and on the maturity of S-NPP products. We compared current requirements to prior iterations to assess how recent changes in capabilities have impacted program goals and objectives. We also interviewed JPSS program officials to discuss S-NPP product development, JPSS system development, and changes in requirements for JPSS satellites. To evaluate the quality of NOAA's program schedule, we used an exposure draft of GAO's Schedule Assessment Guide[1] to assess component contractor schedules as well as the program's schedule risk analysis and interviewed cognizant JPSS program office and contractor officials.

To assess NOAA's plans to address potential gaps in polar satellite data, we compared NOAA's gap mitigation plan and contracted alternatives study against risk mitigation and contingency best practices from GAO and advocated by leading organizations,[2] determined planning shortfalls and key remaining activities for NOAA to accomplish, and interviewed NOAA headquarters and JPSS program officials about their plans.

We conducted this performance audit from October 2012 through September 2013 in accordance with generally accepted government auditing standards. Those standards require that we plan and perform the audit to obtain sufficient, appropriate evidence to provide a reasonable basis for our findings and conclusions based on our audit objectives. Additional details on our objectives, scope, and methodology are provided in appendix I.

BACKGROUND

Since the 1960s, the United States has operated two separate operational polar-orbiting meteorological satellite systems: the Polar-orbiting Operational Environmental Satellite (POES) series, which is managed by NOAA, and the Defense Meteorological Satellite Program (DMSP), which is managed by the Air Force.[3] These satellites obtain environmental data that are processed to provide graphical weather images and specialized weather products. These satellite data are also the predominant input to numerical weather prediction models, which are a primary tool for forecasting weather days in advance—including forecasting the path and intensity of hurricanes. The weather products and models are used to predict the potential impact of severe weather so that communities and emergency managers can help prevent and mitigate its effects. Polar satellites also provide data used to monitor environmental phenomena, such as ozone depletion and drought conditions, as well as data sets that are used by researchers for a variety of studies such as climate monitoring.

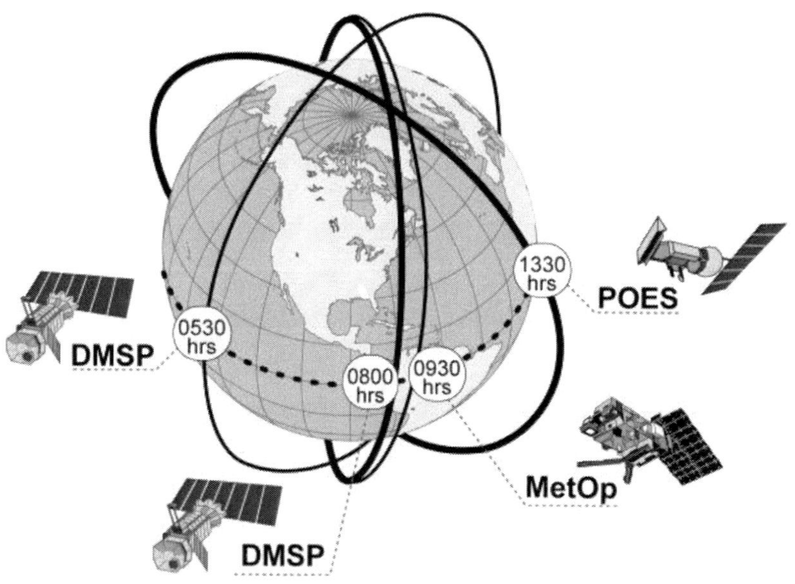

Source: GAO, based on NPOESS Integrated Program Office and DOD data, Map Resources (globe).

Figure 1. Configuration of Operational Polar Satellites.

Unlike geostationary satellites, which maintain a fixed position relative to the earth, polar-orbiting satellites constantly circle the earth in an almost north-south orbit, providing global coverage of conditions that affect the weather and climate. Each satellite makes about 14 orbits a day. As the earth rotates beneath it, each satellite views the entire earth's surface twice a day. Currently, there is one operational POES satellite and two operational DMSP satellites that are positioned so that they cross the equator in the early morning, midmorning, and early afternoon. In addition, the government relies on a European satellite, called the Meteorological Operational (MetOp) satellite, for satellite observations in the midmorning orbit.[4] In addition to the operational satellites, NOAA, the Air Force, and a European weather satellite organization maintain older satellites that still collect some data and are available to provide limited backup to the operational satellites should they degrade or fail. The last POES satellite was launched in February 2009. The Air Force plans to launch its two remaining DMSP satellites as needed. Figure 1 illustrates the current operational polar satellite constellation.

Polar Satellite Data and Products

Polar satellites gather a broad range of data that are transformed into a variety of products. Satellite sensors observe different bands of radiation wavelengths, called channels, which are used for remotely determining information about the earth's atmosphere, land surface, oceans, and the space environment. When first received, satellite data are considered raw data. To make them usable, processing centers format the data so that they are time-sequenced and include earth-location and calibration information. After formatting, these data are called raw data records. The centers further process these raw data records into channel-specific data sets, called sensor data records and temperature data records. These data records are then used to derive weather and climate products called environmental data records. These environmental data records include a wide range of atmospheric products detailing cloud coverage, temperature, humidity, and ozone distribution; land surface products showing snow cover, vegetation, and land use; ocean products depicting sea surface temperatures, sea ice, and wave height; and characterizations of the space environment. Combinations of these data records (raw, sensor, temperature, and environmental data records) are also used to derive more sophisticated products, including outputs from numerical weather models and assessments of climate trends. Figure 2 is a simplified depiction of the various stages of satellite data processing, and figure 3 depicts examples of two different weather products.

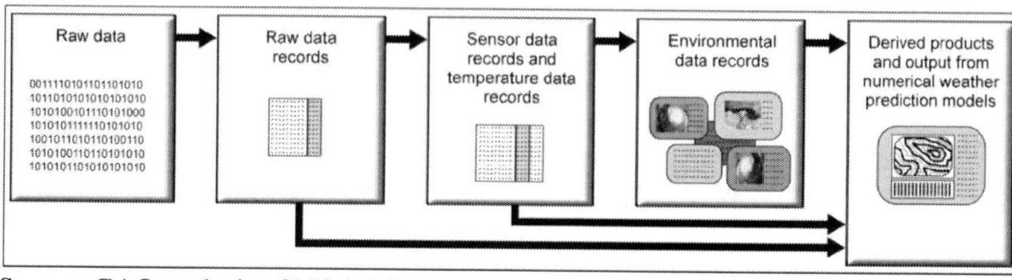

Source: GAO analysis of NOAA information.

Figure 2. Stages of Satellite Data Processing.

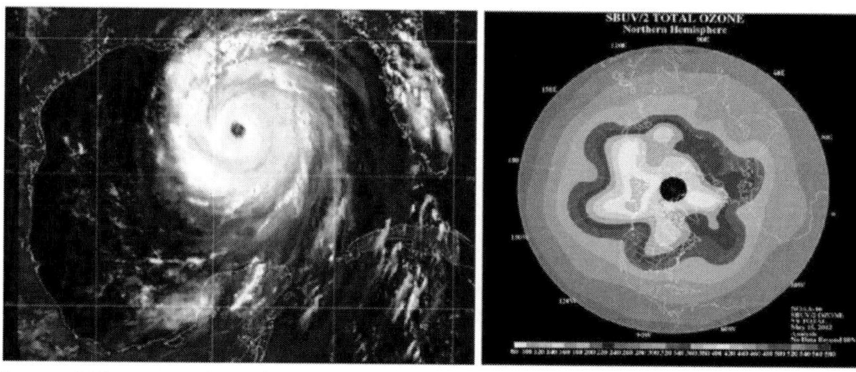

Source: NOAA's National Environmental Satellite Data and Information Service.

Note: The figure on the left is a POES Image of Hurricane Katrina in 2005; the figure on the right is an analysis of ozone concentration produced from POES satellite data.

Figure 3. Examples of Weather Products.

The NPOESS Program: Inception, Challenges, and Divergence

With the expectation that combining the POES and DMSP programs would reduce duplication and result in sizable cost savings, a May 1994 Presidential Decision Directive required NOAA and DOD to converge the two satellite programs into a single satellite program—NPOESS—capable of satisfying both civilian and military requirements.[5] The converged program, NPOESS, was considered critical to the nation's ability to maintain the continuity of data required for weather forecasting and global climate monitoring. NPOESS satellites were expected to replace the POES and DMSP satellites in the morning, midmorning, and afternoon orbits when they neared the end of their expected life spans.

To manage this program, DOD, NOAA, and NASA formed a tri-agency Integrated Program Office, with NOAA responsible for overall program management for the converged system and for satellite operations, the Air Force responsible for acquisition, and NASA responsible for facilitating the development and incorporation of new technologies into the converged system.

When the primary NPOESS contract was awarded in August 2002, the program was estimated to cost about $7 billion through 2018. The program was to include the procurement and launch of 6 satellites over the life of the program, with each satellite hosting a subset of 13 instruments. The planned instruments included 11 environmental sensors, and two systems supporting specific user services (see table 1). To reduce the risk involved in developing new technologies and to maintain climate data continuity, the program planned to launch the demonstration satellite in May 2006.[6] This satellite was intended to demonstrate the functionality of selected instruments that would later be included on the NPOESS satellites. The first NPOESS satellite was to be available for launch in March 2008.

Table 1. Anticipated NPOESS Instruments, as of July 2002

Instrument	Instrument type	Description
Advanced technology microwave sounder (ATMS)	Environmental sensor	Measures microwave energy released and scattered by the atmosphere; to be used in combination with the cross-track infrared sounder to produce daily global atmospheric temperature, humidity, and pressure profiles.
Aerosol polarimetry sensor	Environmental sensor	Retrieves specific aerosol (liquid droplets or solid particles suspended in the atmosphere, such as sea spray, smog, and smoke) and cloud measurements.
Conical microwave imager/sounder	Environmental sensor	Collects microwave images and data needed to measure rain rate, ocean surface wind speed and direction, amount of water in the clouds, and soil moisture, as well as temperature and humidity at different atmospheric levels.

Table 1. (Continued)

Instrument	Instrument type	Description
Cross-track infrared sounder (CrIS)	Environmental sensor	Collects measurements of the infrared radiation emitted and scattered by the Earth and atmosphere to determine the vertical distribution of temperature, moisture, and pressure in the atmosphere.
Data collection system	System providing services to users	Collects environmental data from platforms around the world and delivers them to users worldwide.
Earth radiation budget sensor	Environmental sensor	Measures solar short-wave radiation and long-wave radiation released by the Earth back into space on a worldwide scale to enhance long-term climate studies.
Global positioning system occultation sensor	Environmental sensor	Measures the refraction of radio wave signals from the Global Positioning System and Russia's Global Navigation Satellite System to characterize the ionosphere and information related to the vertical distribution of temperature and moisture of the atmosphere.
Ozone mapper/ profiler suite (OMPS)	Environmental sensor	Collects data needed to measure the amount and distribution of ozone in the Earth's atmosphere. Consists of two components (nadir and limb) that can be provided separately.
Radar altimeter	Environmental sensor	Measures variances in sea surface height/topography and ocean surface roughness, which are used to determine sea surface height, significant wave height, and ocean surface wind speed and to provide critical inputs to ocean forecasting and climate prediction models.
Search and rescue satellite-aided tracking system	System providing services to users	A subsystem that detects and locates aviators, mariners, and land-based users in distress.
Space environmental sensor suite	Environmental sensor	Collects data to identify, reduce, and predict the effects of space weather on technological systems, including satellites and radio links.

Instrument	Instrument type	Description
Total and spectral solar irradiance sensor	Environmental sensor	Monitors and captures total and spectral solar irradiance data.
Visible/infrared imager radiometer suite (VIIRS)	Environmental sensor	Collects images and radiometric data used to provide information on the Earth's clouds, atmosphere, ocean, and land surfaces.

Source: GAO analysis of data from the former NPOESS Integrated Program Office.

In the years after the program was initiated, NPOESS encountered significant technical challenges in sensor development, program cost growth, and schedule delays. By November 2005, we estimated that the program's cost had grown to $10 billion, and the schedule for the first launch was delayed by almost 2 years.[7] These issues led to a 2006 decision to restructure the program, which reduced the program's functionality by decreasing the number of planned satellites from 6 to 4, and the number of instruments from 13 to 9. As part of the decision, officials decided to reduce the number of orbits from three (early morning, midmorning, and afternoon) to two (early morning and afternoon) and to rely solely on the European satellites for midmorning orbit data.

Even after the restructuring, however, the program continued to encounter technical issues in developing two sensors, significant tri-agency management challenges, schedule delays, and further cost increases. Because the schedule delays could lead to satellite data gaps, in March 2009 agency executives decided to use S-NPP as an operational satellite.[8] Later, in August 2009, faced with costs that were expected to reach about $15 billion and launch schedules that were delayed by over 5 years, the Executive Office of the President formed a task force, led by the Office of Science and Technology Policy, to investigate the management and acquisition options that would improve the NPOESS program. As a result of this review, in February 2010, the Director of the Office of Science and Technology Policy announced that NOAA and DOD would no longer jointly procure the NPOESS satellite system; instead each agency would plan and acquire its own satellite system.[9] Specifically, NOAA would be responsible for the afternoon orbit and the observations planned for the first and third satellites. DOD would be responsible for the early morning orbit and the observations planned for the second and fourth satellites. The partnership with the European satellite agencies for the midmorning orbit was to continue as planned. When this decision was announced, NOAA and NASA immediately began planning for a new satellite program in the afternoon orbit called JPSS. DOD began planning for a new satellite program in the morning orbit, called the Defense Weather Satellite System, but later decided to terminate the program and reassess its requirements, as directed by Congress.

Overview of Initial NOAA Plans for the JPSS Program

After the decision was made to disband the NPOESS program in 2010, NOAA began the JPSS satellite program. Key plans included:

- relying on NASA for system acquisition, engineering, and integration;
- completing, launching, and supporting S-NPP;
- acquiring and launching two satellites for the afternoon orbit, called JPSS-1 and JPSS-2;
- developing and integrating five sensors on the two satellites;
- finding alternative host satellites for selected instruments that would not be accommodated on the JPSS satellites; and
- providing ground system support for S-NPP, JPSS, and the Defense Weather Satellite System; data communications for MetOp and DMSP; and data processing for NOAA's use of microwave data from an international satellite.

In 2010, NOAA estimated that the life cycle costs of the JPSS program would be approximately $11.9 billion for a program lasting through fiscal year 2024, which included $2.9 billion in NOAA funds spent on NPOESS through fiscal year 2010.[10] Subsequently, the agency undertook a cost estimating exercise where it validated that the cost of the full set of JPSS functions from fiscal year 2012 through fiscal year 2028 would be $11.3 billion. After adding the agency's sunk costs, which had increased to $3.3 billion through fiscal year 2011, the program's life cycle cost estimate totaled $14.6 billion.[11] This amount was $2.7 billion higher than the $11.9 billion estimate for JPSS when NPOESS was disbanded in 2010.

In working with the Office of Management and Budget to establish the president's fiscal year 2013 budget request, NOAA officials stated that they agreed to cap the JPSS life cycle cost at $12.9 billion through 2028, to fund JPSS at roughly $900 million per year through 2017, and to merge funding for two climate sensors into the JPSS budget. Because this cap was $1.7 billion below the expected $14.6 billion life cycle cost of the full program, NOAA decided to remove selected elements from the satellite program. Table 2 compares the planned cost, schedule, and scope of NOAA's satellite programs at different points in time.

Table 2. A Comparison of NPOESS and JPSS, at Different Points in Time

Key area	NPOESS program before it was restructured (as of May 2006)	NPOESS program after it was restructured (as of June 2006)	NPOESS program prior to being disbanded (as of February 2010)	JPSS program (as of May 2010)	JPSS program (as of June 2012)
Life cycle range	1995-2020	1995-2026	1995-2026	1995-2024	1995-2028
Estimated life cycle cost	$8.4 billion	$12.5 billion	$13.95+ billion[a]	$11.9 billion (which includes about $2.9 billion spent through fiscal year 2010 on NPOESS)	$12.9 billion (which includes about $3.3 billion spent through fiscal year 2011 on NPOESS and JPSS)
Number of satellites	6 (in addition to S-NPP)	4 (in addition to S-NPP)	4 (in addition to S-NPP)	2 (in addition to S-NPP)	2 (in addition to S-NPP)
Number of orbits	3 (early morning, midmorning, afternoon)	2 (early morning and afternoon; would rely on European satellites for midmorning orbit data)	2 (early morning and afternoon; would rely on European satellites for midmorning orbit data)	1 (afternoon orbit) (DOD and European satellites would provide early and midmorning orbits, respectively)	1 (afternoon orbit) (DOD and European satellites would provide early and midmorning orbits, respectively)
Launch schedule	S-NPP by October 2006 First NPOESS (C1) by November 2009 Second NPOESS (C2) by June 2011	S-NPP by January 2010 C1 by January 2013 C2 by January 2016 C3 by January 2018 C4 by January 2020	S-NPP no earlier than September 2011 C1 by March 2014 C2 by May 2016 C3 by January 2018 C4 by January 2020	S-NPP—no earlier than September 2011 JPSS-1 available in 2015 JPSS-2 available in 2018	S-NPP—successfully launched in October 2011 JPSS-1 by March 2017 JPSS-2 by December 2022

Table 2. (Continued)

Key area	NPOESS program before it was restructured (as of May 2006)	NPOESS program after it was restructured (as of June 2006)	NPOESS program prior to being disbanded (as of February 2010)	JPSS program (as of May 2010)	JPSS program (as of June 2012)
Life cycle range	1995-2020	1995-2026	1995-2026	1995-2024	1995-2028
Number of sensors	11 sensors and 2 user services systems	S-NPP: 4 sensors C1: 6 sensors C2: 2 sensors C3: 6 sensors C4: 2 sensors	S-NPP: 5 sensors C1: 7 sensors[b] C2: 2 sensors C3: 6 sensors C4: 2 sensors	S-NPP: 5 sensors JPSS-1: 5 sensors[c] JPSS-2: 5 sensors	S-NPP: 5 sensors JPSS-1: 5 sensors[c] JPSS-2: 5 sensors Free flyer-1: 3 sensors[d] Free flyer-2: 3 sensors

Source: GAO analysis of NOAA, DOD, and task force data.

[a] Although the program baseline was $13.95 billion in February 2010, we estimated in June 2009 that this cost could grow by about $1 billion. In addition, officials from the Executive Office of the President stated that they reviewed life cycle cost estimates from DOD and the NPOESS program office of $15.1 billion and $16.45 billion, respectively.

[b] In May 2008, the NPOESS Executive Committee approved an additional sensor—the Total and Spectral Solar Irradiance Sensor—for the C1 satellite.

[c] The five sensors are ATMS, the Clouds and the Earth's Radiant Energy System (CERES), CrIS, OMPS, and VIIRS. NOAA also committed to finding an alternative spacecraft and launch accommodation for the Total and Spectral Solar Irradiance Sensor, the Advanced Data Collection System, and the Search and Rescue Satellite-Aided Tracking system.

[d] NOAA planned to launch two stand-alone satellites, called free flyer satellites, to accommodate the Total and Spectral Solar Irradiance Sensor, Search and Rescue Satellite-Aided Tracking system, and an Advanced Data Collection System.

Prior GAO Work Recommended Actions to Address the Risk of Gaps in Polar Satellite Data

We have issued a series of reports on the NPOESS and JPSS programs highlighting technical issues, cost growth, and key management challenges affecting the tri-agency program structure.[12] In June 2012, we reported that while NOAA officials communicated publicly and often about the risk of a polar satellite data gap, the agency had not established plans to mitigate the gap.[13] At the time, NOAA officials stated that the agency would continue to use existing satellites as long as they provide data and that there were no viable alternatives to the JPSS program. However, our report noted that a more comprehensive mitigation plan was essential since it is possible that other governmental, commercial, or foreign satellites could supplement the polar satellite data. Because it could take time to adapt ground systems to receive, process, and disseminate an alternative satellite's data, we noted that any delays in establishing mitigation plans could leave the agency little time to leverage its alternatives. We recommended that NOAA establish mitigation plans for risks associated with pending satellite gaps in the afternoon orbit as well as potential gaps in the early morning and midmorning orbits. NOAA agreed with the report's recommendation and noted that the National Environmental Satellite, Data, and Information Service—a NOAA component agency—had performed analyses on how to mitigate potential gaps in satellite data and planned to provide a report by August 2012.

More recently, in February 2013, we added the potential gap in weather satellite data to our biennial High-Risk list.[14] In that report, we noted that satellite data gaps in the morning or afternoon polar orbits would lead to less accurate and timely weather forecasting; as a result, advanced warning of extreme events would be affected. Such extreme events could include hurricanes, storm surges, and floods. For example, the National Weather Service performed case studies to demonstrate how its forecasts would have been affected if there were no polar satellite data in the afternoon orbit, and noted that its forecasts for the "Snowmaggedon" winter storm that hit the Mid-Atlantic coast in February 2010 would have predicted a less intense storm further east, with about half of the precipitation at 3, 4, and 5 days before the event. Specifically, the models would have under-forecasted the amount of snow by at least 10 inches. Similarly, a European weather organization[15] recently reported that NOAA's forecasts of Hurricane Sandy's track could have been hundreds of miles off without polar-orbiting satellites—rather than identifying the New Jersey landfall within 30 miles 4 days before landfall, the models would have shown the storm remaining at sea. Such degradation in forecasts and warnings would place lives, property, and our nation's critical infrastructure in danger.

We reported that the length of an afternoon polar satellite data gap could span from 17 months to 3 years or more. In one scenario, S-NPP would last its full expected 5-year life (to October 2016), and JPSS-1 would launch as soon as possible (in March 2017) and undergo on-orbit checkout for a year (until March 2018). In that case, the data gap would extend 17 months. In another scenario, S-NPP would last only 3 years as noted by NASA managers concerned with the workmanship of selected S-NPP sensors. Assuming that the JPSS-1 launch occurred in March 2017 and the satellite data were certified for official use by March 2018, this gap would extend for 41 months. Of course, any problems with JPSS1 development could delay the launch date and extend the gap period. Figure 4 depicts four possible gap scenarios.

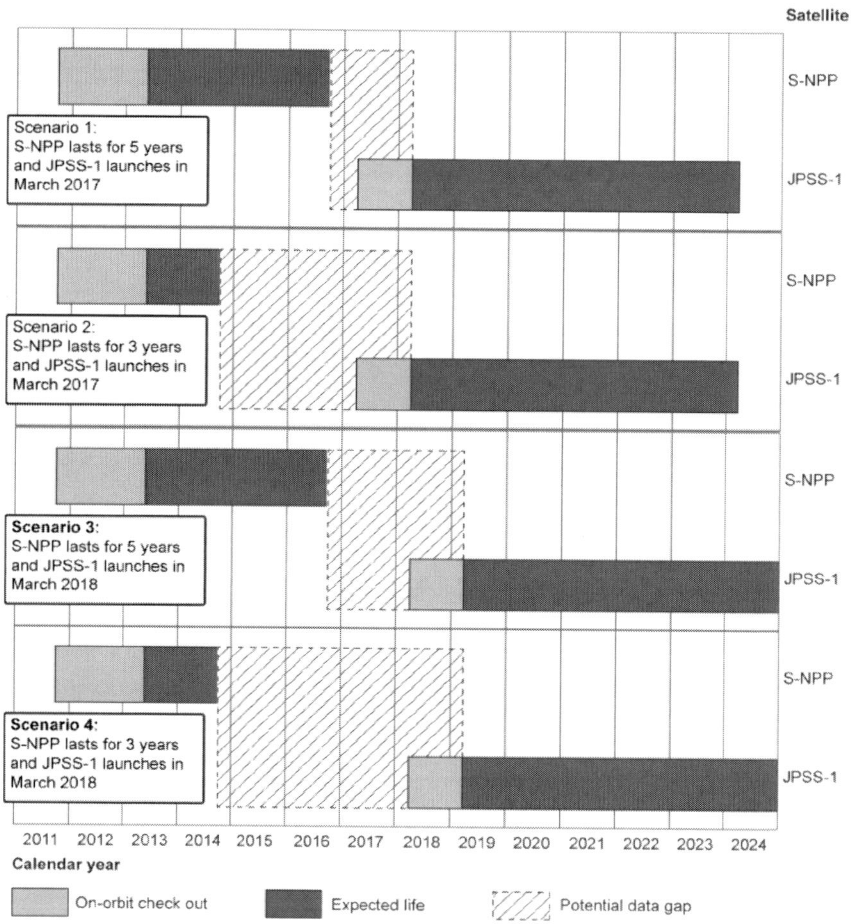

Source: GAO analysis of NOAA data.

Figure 4. Potential Gaps in Polar Satellite Data in the Afternoon Orbit.

We also noted that NOAA had recently established a mitigation plan for a potential 14- to 18-month gap in the afternoon orbit, which identified and prioritized options for obtaining critical observations, including alternative satellite data sources and improvements to data assimilation in models and listed technical, programmatic, and management steps needed to implement these options. However, these plans were only a beginning. We suggested that NOAA must make difficult decisions on which steps it would implement to ensure that its mitigation plans are viable when needed, including how these plans would be integrated with the agency's broader end-to-end plans for sustaining weather forecasting capabilities.

NOAA Has Made Progress on JPSS Development, but Continues to Face Challenges in Completing S-NPP Products, Revising the Program's Scope, and Meeting Schedules

NOAA has made progress towards JPSS program objectives of sustaining the continuity of NOAA's polar-orbiting satellite capabilities through the S-NPP, JPSS-1, and JPSS-2 satellites by (1) delivering S-NPP data to weather forecasters and (2) by completing significant instrument and spacecraft development for the JPSS-1 satellite. However, the program is behind schedule in validating the readiness of S-NPP products and has experienced delays on the ground system schedules for the JPSS-1 satellite. Moreover, the program is moving to revise its scope and objectives to reduce costs and prioritize NOAA's weather mission. Until it addresses challenges in product and ground system development, the program office may continue to experience delays in delivering actionable S-NPP data to users and in meeting program development schedules.

Weather Forecasters Are Using Selected S-NPP Products, but the JPSS Program Is Behind Schedule in Validating Products and Unaware of the Full Extent to Which They Are Being Used

In order to sustain polar-orbiting earth observation capabilities through the S-NPP satellite, over the past 18 months the JPSS program had planned to complete activation and commissioning of the S-NPP satellite, transition the satellite from interim to routine operations, and deliver 76 data products that were precise enough for use in operational weather observations and forecasts. To develop the precise data products, NOAA established a process for calibrating and validating its products. Under this process, most products (which are primarily sensor data records and environmental data records) proceed through three different levels of algorithm maturity—the beta, provisional, and validated levels.[16] NOAA had originally planned to complete efforts to validate S-NPP products by October 2013, which was 2 years after the S-NPP satellite was launched. It is not enough, however, to simply deliver validated products. Both the Software Engineering Institute and GAO recommend tracking whether customers are receiving the expected value from products once they are deployed, and whether corrective actions are needed.[17] Moreover, in April 2013 the Executive Office of the President's National Science and Technology Council released a national strategy for civil earth observations that called for agencies to, among other things, track the extent to which earth observation data are actually being used, track whether the data had an impact, and provide data users a mechanism to provide feedback regarding ease of use, suspected quality issues, and other aspects of the data.[18]

The JPSS program has made progress on S-NPP since launching the satellite in October 2011. Specifically, the program completed satellite activation and commissioning in March 2012, and transitioned from interim operations under NASA to routine operations under NOAA in February 2013. The program also made key upgrades to the ground system supporting S-NPP. For example, in November 2012 the office completed an interim backup command and control facility that could protect the health and safety of the satellite if unexpected issues occurred at the primary mission operations facility. In addition, the JPSS program office has been working to calibrate and validate S-NPP products in order to make them precise enough for use in weather-related operations. While the program office plans to have 18 products validated for operational use by September 2013, it is behind schedule for the other products. Specifically, the program expects to complete validating 35 S-NPP products by the end of September 2014 and 1 other product by the end of September 2015, almost 1 and 2 years later than originally planned. In addition, the program office reported that 15 products do not need to be validated, one product's validation date has not been established, and 6 products do not have estimated validation dates because the program plans to remove them from its requirements. The program categorized its products by their priority, ranging from priority-1 for the highest priority products, to priority-4 for the lowest priority products.

Table 3. Estimated Completion Dates for S-NPP Products, as of July 2013

Number of products	Priority level (number of products)	Examples of products	Validated by
18	Priority 1 or 2: 4	ATMS land surface emissivity Advanced Microwave Scanning Radiometer-2 sensor data record VIIRS polar winds VIIRS vegetation fraction	End of September 2013
	Priority 3 or 4: 14	ATMS rainfall rate OMPS-nadir profile ozone OMPS-nadir ozone total column Advanced Microwave Scanning Radiometer-2 cloud liquid water	
35	Priority 1 or 2: 5	ATMS sensor data record CrIS sensor data record VIIRS sensor data record VIIRS imagery	End of September 2014
	Priority 3 or 4: 29	OMPS-nadir sensor data record VIIRS active fires VIIRS cloud optical thickness CrIS infrared ozone profile	
	No priority: 1	VIIRS ocean color/chlorophyll	
1	Priority 3 or 4: 1	VIIRS vegetation health index suite	End of September 2015
15	Priority not assigned	Application packets and raw data records	Not applicable[a]
1	Priority 1 or 2: 1	Advanced Microwave Scanning Radiometer-2 temperature data record	Date is still to be determined
6	Priority 1 or 2: 0	Not applicable	Planned to be removed from requirements[b]
	Priority 3 or 4: 5	CERES sensor data record	
	No priority: 1	OMPS-limb profiler sensor data record	

Source: GAO analysis of NOAA documents.

Notes:

[a]Program officials stated that selected products, including raw data records and application packets, do not undergo validation.

[b]Program officials expect that these requirements will be removed once the transfer of instruments is approved through the fiscal year 2014 budget process.

According to NOAA and NASA officials, the S-NPP products' validation has been delayed in part because of issues initially identified on VIIRS that had to be corrected and additional time needed to validate environmental data record products that require observations of seasonal weather phenomena. Further, program officials stated that they rebaselined the planned product validation timelines in November 2011 and have been generally meeting the target dates of this revised plan. Table 3 illustrates program-reported data on the number of products in each priority level, examples of products, and the estimated validation date for the last product at each level.

Even though S-NPP products are not at the validated stage in which products are ready for operational use, the National Weather Service (NWS) has accepted certain products for use in its operational systems. For example, the JPSS program office reported that NWS is using ATMS temperature data records in its operational forecasts, and that the Alaska Weather Forecast Offices are using VIIRS imagery in its forecasts. In addition, NWS's National Centers for Environmental Prediction is evaluating CrIS sensor data records for use in numerical weather prediction, but has not yet used the data operationally because it is in the midst of a computer upgrade. Officials also stated that the program obtains information on the operational use of S-NPP data from other NOAA offices, including the National Ocean Service and the National Marine and Fisheries Service.

While NOAA is aware of these uses, it does not track the extent to which key satellite data users—including users from the Air Force, Navy, Forest Service, European weather offices, and academic institutions—have incorporated S-NPP data into their operations or if corrective actions are needed to make the products more accurate or more effective for the specific users. Program officials noted that they are not required to tailor products to meet non-NOAA user requirements, and that they do not have a tracking mechanism that would allow them to identify which entities are using the data. They noted, however, that the program obtains informal reports from customer representatives through various working groups and forums, such as the Low-earth Orbiting Requirements Working Group and the JPSS Customer Forum. While these efforts obtain information from known customer groups, they do not meet best practices for actively tracking whether customers are using the products, receiving the expected value, or in need of product corrections. Until the program office tracks the use of S-NPP and future JPSS products, it will not have full knowledge of the extent to which products are being used to assess the value they provide to end users and whether corrective actions are needed. More significantly, without information on who is using S-NPP data, NOAA will be unable to ensure that the significant investment made on this satellite is not wasted.

Development of JPSS Flight Project Is on Track, but Scheduling Issues on the Ground System Have Caused Delays

In order to sustain polar-orbiting earth observation capabilities, the program is working to complete development of the JPSS-1 systems in preparation for a March 2017 launch date. To manage this initiative, the program office organized its responsibilities into two separate projects: (1) the flight project, which includes sensors, spacecraft, and launch vehicles and (2) the ground project, which includes ground-based data processing and command and control systems. Table 4 shows the JPSS projects and their key components.

Table 4. JPSS Projects and Components

Project	Key components and responsibilities
Flight	Sensors: ATMS, CERES, CrIS, OMPS, VIIRS Spacecraft Launch vehicle
Ground	Satellite command, control, and communications Interface data processing segment

Source: NOAA's JPSS program office.

JPSS projects and components are at various stages of system development. The flight project has nearly completed instrument hardware development for the JPSS-1 satellite and has begun testing certain instruments. Also, the flight project completed a major design review for the JPSS-1 satellite's spacecraft. While the flight project's development is on track, the ground project experienced delays in its planned schedule that could further delay major program milestones, including key reviews required to establish the program's cost and schedule baseline.

Flight Project: JPSS-1 Instrument and Spacecraft Development Is On Track

The flight project is generally on track with respect to planned JPSS-1 instrument and spacecraft development efforts. According to program reports of instrument development, the instruments for the JPSS-1 satellite are nearly complete. Specifically, as of July 2013, the instrument hardware ranged from 80 to 100 percent complete. Also, all of the instruments have completed or are scheduled to complete environmental testing reviews in 2013 and are to be delivered to the spacecraft by 2014. The spacecraft completed its critical design review—which evaluates whether the design is appropriately mature to continue with the final design and fabrication—in January 2013.

While individual instruments have experienced delays, the key testing milestones and delivery dates for the instruments and spacecraft have generally held constant since the last key decision point in July 2012. CERES experienced a 10-month slip in its delivery date due to a technical issue with the instrument's internal calibration monitor, and ATMS experienced an 8-month slip to its pre-environmental review due to an issue in one of the sensor's channels, but even accounting for these slips, the instruments have a schedule reserve of 14 and 10 months, respectively. VIIRS is expected to be the last instrument to be delivered to the spacecraft and has a schedule reserve of 6 months. Also, between July 2012 and December 2012 instrument contractors' estimated costs at completion increased by $29 million for ATMS, CrIS, and OMPS, while the cost for VIIRS decreased by $46 million. In addition, based on program reports of technical performance, the instruments and the spacecraft are generally meeting expected technical performance. Table 5 describes the current status of the components of the JPSS-1 flight project.

Table 5. Status of Key Components of the Flight Project Supporting the JPSS-1 Satellite, as of July 2013

Component	Status
Instrument	
ATMS	The instrument completed its hardware development and is in environmental testing. The instrument experienced a technical issue beginning in November 2012 in which science counts in one of the channels were lower than expected. The program replaced channel components with spares and performed additional regression testing. This issue delayed the planned pre-environmental review by 6 months from November 2012 to May 2013. ATMS is expected to be delivered for integration on the spacecraft in March 2014.
CERES	The instrument completed its hardware development and is in environmental testing. The instrument's internal calibration monitor exhibited unstable performance during calibration, which delayed its delivery by 13 months. The cause of this issue is still unknown. CERES underwent pre-environmental review in February 2012. The program office plans to perform additional calibration testing, vacuum performance testing, and regression testing prior to its delivery in October 2013.
CrIS	The instrument completed 80 percent of its hardware development and is in subsystem integration. The instrument's electronic components have been experiencing a power-up issue, possibly due to a timing issue with the digital power supply. The program office replaced the power supply with a spare and is working on resolving the issue and completing acceptance testing.

Component	Status
	Additional work remaining on the instrument includes completing the subsystems, integrating the complete instrument, conducting the pre-environmental review in September 2013, and completing its environmental test program. The expected delivery date for the instrument is August 2014.
OMPS-Nadir	The instrument completed its hardware development and is in environmental testing. The instrument's diffuser experienced degradation during calibration and the adhesive was deemed the root cause. There was no performance impact but the instrument's delivery date slipped 3 months. The program office planned to complete cleaning, regression testing, and preparation of the nadir unit for testing. The instrument completed a pre-environmental review in April 2013, is conducting environmental testing, and is expected to be delivered in August 2014.
VIIRS	The instrument completed 80 percent of its hardware development and is in subsystem integration. Several sensor components have been delivered, installed, and integrated. VIIRS is expected to undergo pre-environmental review in October 2013 with an expected delivery in October 2014.
Spacecraft	The spacecraft completed its critical design review in December 2012; hardware development is ongoing. The spacecraft has an expected delivery in October 2014.
Launch vehicle	NASA awarded a contract for launch services in July 2012. Previously accepted risks for the launch program on S-NPP are being reviewed for applicability and potential mitigation.

Source: GAO analysis of JPSS program office data.

Ground Project: Progress Made, but Facility Scheduling Problems Have Caused Delays

The JPSS ground project has made progress in developing the ground system components, but scheduling issues have caused delays in the deployment of system upgrades. Specifically, between August 2012 and February 2013, the program office defined the ground system's technical performance baseline, ordered and received the first increment of hardware for the next major software release, and transitioned S-NPP operational management from the JPSS program to NOAA's office responsible for satellite operations.

However, the program has delayed the delivery of key ground system upgrades needed to support JPSS-1 because the facilities needed for hardware installation, software development, and testing activities were not available when needed. The ground system upgrades, called block 1.5 and 2.0, were originally scheduled to be delivered in January and December 2015, respectively. To address the problem in scheduling the facilities, NOAA delayed the delivery of block 1.5 and merged it with block 2.0. The program is now expecting to deliver both upgrades in December 2015. We have previously reported that compressing system development schedules introduces program risk because it implies the need to accomplish a larger number of activities in parallel and on time before the next major event can occur as planned.[19] As a result, any complications in the merged ground system upgrades could affect the system's readiness to support the JPSS-1 launch date.

NOAA Revised Program Scope to Focus on Weather Priorities and Reduce Costs

While NOAA is moving forward to complete product development on the S-NPP satellite and system development on the JPSS-1 satellite, the agency recently made major revisions to the program's scope and planned capabilities and is moving to implement other scope changes as it finalizes its plans pending congressional approval. We previously reported that, as part of its fiscal year 2013 budget process, NOAA was considering removing selected elements of the program in order to reduce total program costs from $14.6 billion to $12.9 billion.[20] By October 2012, NOAA made the following changes in the program's scope:

- develop two (instead of three) TSIS instruments as well as two free-flyer spacecraft and launch vehicles to accommodate the instruments;
- reduce the previously planned network of fifteen ground-based receptor stations to two receptor sites at the north pole and two sites at the south pole;
- increase the time it takes to obtain satellite data and deliver it to the end user from 30 minutes to 80 minutes on the JPSS-2 satellite;[21]
- not install an interface data processing segment at the two Navy locations or at the Air Force Weather Agency; and
- withdraw future support for ground operations for DOD's Defense Weather Satellite System, which was subsequently cancelled.

More recently, as proposed by the administration, NOAA began implementing additional changes in the program's scope and objectives in order to meet the agency's highest-priority needs for weather forecasting and reduce program costs from $12.9 billion to $11.3 billion. Specifically, NOAA has begun to:

- Transfer requirements for building the OMPS-limb and CERES follow-on climate sensors for the JPSS-2 satellite to NASA.

- Transfer the first free-flyer mission from the JPSS program to a separate NOAA program, called the Polar Free Flyer program, and cancel the second free-flyer mission. More information on the Polar Free Flyer program is provided in appendix II.
- Eliminate requirements for a legacy type of broadcast transmitter, which, according to NOAA officials, is in a spectrum range being crowded out by terrestrial users and is consistent with its European partners' plans.
- Reduce science and algorithm requirements for lower-priority data products.
- Reduce operations and sustainment costs based on increased efficiencies through moving from customized components to more off the-shelf solutions.
- Reduce the mission life cycle by 3 years from 2028 to 2025.

Table 6. Reported Cost Reductions Associated with JPSS Changes

Timing of revisions	Program reported cost reductions	Major scope changes and other revisions
June 2012	$1.7 billion through 2028 (from the $14.6 billion to $12.9 billion life cycle cost estimates)	Restructured free flyer missions ($800M) Revised operations and sustainment concept ($700M) Reduced reserve estimates ($200M)
April 2013	$1.6 billion through 2025 (from the $12.9 billion to $11.3 billion life cycle cost estimates)	Transferred first free flyer mission to Polar Free Flyer program and transferred a TSIS, CERES, and OMPS-Limb instrument to NASA ($750M) Reduced program lifetime by 3 years, from 2028 to 2025 ($390M) Eliminated second free-flyer mission and one type of communication downlink ($240M) Lowered expected costs for the JPSS-1 launch vehicle and launch services ($9M) Saved in other areas, including costs saved by acquiring VIIRS spare parts from the Air Force and eliminating enhanced data processing of data obtained from the Global Change Observation Mission-Water satellite ($211M)

Source: GAO analysis of JPSS program data.

While we were unable to precisely itemize the reductions in costs associated with various program changes, program officials provided rough estimates. The following table summarizes the reported cost reductions associated with key changes to the JPSS program.

While there are a number of reasons for individual changes in the program, the key reason for the June 2012 changes was to meet the program's $12.9 billion cost cap. The reasons for the more recent changes were to reduce mission costs and complexity, focus JPSS priorities on NOAA's weather forecasting mission, and identify opportunities to reduce potential gaps between JPSS satellites, all of which an independent study on NOAA's satellite program recommended in July 2012.

While these are worthy goals, the changes NOAA implemented over the last 2 years will have an impact on those who rely on polar satellite data. Specifically, satellite data products will be delivered more slowly than anticipated because of the reduction in the number of ground stations, and military users may not obtain the variety of products once anticipated at the rates anticipated because of the removal of their ground-based processing subsystems. Further, while not as obvious, the impact of other changes, including the removal of the communications downlink and the reduction of requirements for certain algorithms, could also affect specific groups of satellite data users. As NOAA moves to implement these program changes, it will be important to assess and understand the impact the changes will have on satellite data users.

JPSS Schedules Demonstrate Multiple Best Scheduling Practices, but Integration Problems and Other Weaknesses Reduce Confidence in the JPSS-1 Launch Date

The JPSS program office has established a preliminary integrated master schedule and implemented multiple scheduling best practices, but the integrated master schedule is not complete and weaknesses in component schedules significantly reduce the program's schedule quality as well as management's ability to monitor, manage, and forecast satellite launch dates. The incomplete integrated master schedule and shortfalls in component schedules are due in part to the program's plans to further refine the schedule as well as schedule management and reporting requirements that varied among contractors. Further, while the program is reporting a 70 percent confidence level in the JPSS-1 launch date, its analysis is likely to be overly optimistic because it was not conducted with an integrated schedule and included a component schedule with weaknesses. Until the program office completes its integrated master schedule and addresses weaknesses in component schedules, it will lack the information it needs to effectively monitor development progress, manage dependencies between schedules, and forecast the JPSS-1 satellite's completion and launch.

The JPSS Program Has Not Yet Established a Complete Integrated Master Schedule

According to our guidance on best practices in scheduling,[22] the success of a program depends in part on having an integrated and reliable master schedule that defines when and how long work will occur and how each activity is related to the others. The program schedule provides not only a road map for systematic project execution but also the means by which to gauge progress, identify and resolve potential problems, and promote accountability at all levels of the program. An integrated master schedule constitutes a program schedule as a network of logically linked sequences of activities that includes the entire required scope of effort, including the effort necessary from the government, contractors, and other key parties for a program's successful execution from start to finish. Although the integrated master schedule includes all government, contractor, and external effort, the government program management office is ultimately responsible for its development and maintenance.

The JPSS program office provided a preliminary integrated master schedule in June 2013, but this schedule is incomplete. The program's June 2013 schedule is its first attempt to document a programwide integrated master schedule since it began in October 2010. The schedule contains the scope of work for key program components, such as the JPSS-1 and JPSS-2 satellites and the ground system, and cites linkages to more detailed component schedules. However, significant weaknesses exist in the program's schedule. Specifically, about one-third of the schedule is missing logical relationships called dependencies that are needed to depict the sequence in which activities occur. Because a logic relationship dictates the effect of an on-time, delayed, or accelerated activity on subsequent activities, any missing or incorrect logic relationship is potentially damaging to the entire network. Complete network logic between all activities is essential if the schedule is to correctly forecast the start and end dates of activities within the plan. Program documentation acknowledges that this schedule is not yet complete and the program office plans to refine it over time. Until the program office completes its integrated schedule and includes logically linked sequences of activities, it will lack the information it needs to effectively monitor development progress, manage dependencies, and forecast the JPSS-1 satellite's completion and launch.

The Quality of JPSS-1 Component Schedules Is Inconsistent

Our scheduling guidance identifies ten best practices that support four characteristics of a high-quality, reliable schedule—comprehensive, well-constructed, credible, and controlled.[23] A *comprehensive* schedule includes all government and contractor activities, reflects resources (labor, materials, and overhead) needed to do the work, and realistically reflects how long each activity will take. A *well-constructed* schedule includes activities that are sequenced with the most straightforward logic possible, a critical path[24] that represents a true model of the activities that drive the project's earliest completion date, and total float that accurately depicts schedule flexibility. A *credible* schedule reflects the order of events necessary to achieve aggregated products or outcomes (horizontal traceability) and maps varying levels of the schedule to one another (vertical traceability). Also, a credible schedule includes data about risks and opportunities that are used to predict a level of confidence in meeting the project's completion date. A *controlled* schedule is updated periodically by trained schedulers using actual progress and logic to realistically forecast dates for program activities and is compared against a designated baseline schedule to measure, monitor, and report the project's progress. The JPSS program office is applying NASA's schedule management handbook guidance to manage its schedules, which is largely consistent with our guidance on scheduling best practices. Table 7 provides more detail on the best practices and key characteristics of a reliable schedule.

The quality of three selected component schedules supporting the JPSS1 mission—VIIRS, the spacecraft, and the ground system—was inconsistent with respect to implementing the characteristics of a high-quality, reliable schedule.[25] Each schedule had strengths and weaknesses with respect to sound scheduling practices, but VIIRS was a stronger schedule with fewer weaknesses compared to the ground system and spacecraft schedules. Since the reliability of an integrated schedule depends in part on the reliability of its subordinate schedules, schedule quality weaknesses in these schedules could transfer to an IMS derived from them. Table 8 identifies the quality of each of the selected JPSS-1 component schedules based on the extent to which they met ten best practices of high-quality and reliable schedules; the discussion that follows highlights observed strengths and weaknesses from each schedule. In addition, appendix III includes a more detailed assessment of each schedule against the ten best practices.

Table 7. The Four Characteristics and Ten Best Practices of a High-Quality and Reliable Schedule

Characteristic	Best practice	Description
Comprehensive	Capturing all activities	The schedule should reflect all activities as defined in the project's work breakdown structure, which defines in detail the work necessary to accomplish a project's objectives, including activities both the owner and contractors are to perform.
	Assigning resources to all activities	The schedule should reflect the resources (labor, materials, overhead) needed to do the work, whether they will be available when needed, and any funding or time constraints.
	Establishing the duration of all activities	The schedule should realistically reflect how long each activity will take. When the duration of each activity is determined, the same rationale, historical data, and assumptions used for cost estimating should be used. Durations should be reasonably short and meaningful and allow for discrete progress measurement. Schedules that contain planning and summary planning packages as activities will normally reflect longer durations until broken into work packages or specific activities.
Well-constructed	Sequencing all activities	The schedule should be planned so that critical project dates can be met. To do this, activities need to be logically sequenced—that is, listed in the order in which they are to be carried out. In particular, activities that must be completed before other activities can begin (predecessor activities), as well as activities that cannot begin until other activities are completed (successor activities), should be identified. Date constraints and lags should be minimized and justified. This helps ensure that the interdependence of activities that collectively lead to the completion of events or milestones can be established and used to guide work and measure progress.

Table 7. (Continued)

Characteristic	Best practice	Description
	Confirming that the critical path is valid	The schedule should identify the program critical path—the path of longest duration through the sequence of activities. Establishing a valid critical path is necessary for examining the effects of any activity's slipping along this path. The program critical path determines the program's earliest completion date and focuses the team's energy and management's attention on the activities that will lead to the project's success.
	Ensuring reasonable total float	The schedule should identify reasonable float (or slack)—the amount of time by which a predecessor activity can slip before the delay affects the program's estimated finish date—so that the schedule's flexibility can be determined. Large total float on an activity or path indicates that the activity or path can be delayed without jeopardizing the finish date. The length of delay that can be accommodated without the finish date's slipping depends on a variety of factors, including the number of date constraints within the schedule and the amount of uncertainty in the duration estimates, but the activity's total float provides a reasonable estimate of this value. As a general rule, activities along the critical path have the least float.
Credible	Verifying that the schedule can be traced horizontally and vertically	The detailed schedule should be horizontally traceable, meaning that it should link products and outcomes associated with other sequenced activities. These links are commonly referred to as "handoffs" and serve to verify that activities are arranged in the right order for achieving aggregated products or outcomes. The integrated master schedule should also be vertically traceable—that is, varying levels of activities and supporting subactivities can be traced. Such mapping or alignment of levels enables different groups to work to the same master schedule.
	Conducting a schedule risk analysis	A schedule risk analysis uses a good critical path method schedule and data about project schedule risks and opportunities as well as statistical simulation to predict the level of confidence in meeting a program's completion date, determine the time contingency needed for a level of confidence, and identify high-priority risks and opportunities. As a result, the baseline schedule should include a buffer or reserve of extra time.

Characteristic	Best practice	Description
Controlled	Updating the schedule using actual progress and logic	Progress updates and logic provide a realistic forecast of start and completion dates for program activities. Maintaining the integrity of the schedule logic at regular intervals is necessary to reflect the true status of the program. To ensure that the schedule is properly updated, people responsible for the updating should be trained in critical path method scheduling.
	Maintaining a baseline schedule	A baseline schedule is the basis for managing the project scope, the time period for accomplishing it, and the required resources. The baseline schedule is designated the target schedule, subject to a configuration management control process, against which project performance can be measured, monitored, and reported. The schedule should be continually monitored so as to reveal when forecasted completion dates differ from planned dates and whether schedule variances will affect downstream work. A corresponding baseline document explains the overall approach to the project, defines custom fields in the schedule file, details ground rules and assumptions used in developing the schedule, and justifies constraints, lags, long activity durations, and any other unique features of the schedule.

Source: GAO Schedule Assessment Guide, GAO-12-120G (exposure draft).

Table 8. Assessment of JPSS-1 Component Schedule Quality

Schedule characteristic or best practice	Ground system	Spacecraft	VIIRS
Comprehensive			
Capturing all activities	◕	◑	◕
Assigning resources to all activities	◕	◕	◕
Establishing the duration of all activities	◕	◑	●
Well-constructed			
Sequencing all activities	◕	◑	◕
Confirming that the critical path is valid	◔	◑	◕
Ensuring reasonable total float	◑	◑	◕
Credible			
Verifying that the schedule can be traced horizontally and vertically	◑	◑	◕
Conducting a schedule risk analysis	◔	◑	◑
Controlled			

Table 8. (Continued)

Schedule characteristic or best practice	Ground system	Spacecraft	VIIRS
Updating the schedule using actual progress and logic	◕	◕	●
Maintaining a baseline schedule	◕	◑	◕

Source: GAO analysis of detailed schedules and related documentation for the VIIRS instrument, spacecraft, and ground system.

●=Met: The program office or contractor provided complete evidence that satisfies the entire criterion.

◕=Substantially met: The program office or contractor provided evidence that satisfies a large portion of the criterion.

◑=Partially met: The program office or contractor provided evidence that satisfies about half of the criterion.

◔=Minimally met: The program office or contractor provided evidence that satisfies a small portion of the criterion.

○=Not met: The program office or contractor provided no evidence that satisfies any of the criterion.

Ground Schedule

Of the ten best practices, the ground system schedule minimally met two best practices, partially met two best practices, and substantially met six best practices. There were strengths in the ground schedule in that the contractor established a clear process for integrated information between the schedule and its resource management software and the contractor has performed resource leveling on the schedule. In addition, the contractor stated that people responsible for the activities estimated activity durations. Also, the contractor stated that it performs wellness checks on the quality of the schedule after each update to identify issues associated with missing logic or date constraints and provides a monthly status briefing to the JPSS program office that addresses the status of external schedule handoffs.

However, there were also weaknesses in the ground schedule. For example, activities on the critical path with date constraints are preventing accurate calculations of the schedule's total float, or flexibility. In order for the critical path to be valid, the activities on the critical path must also have reasonable total float. Without a critical path that accurately calculates schedule flexibility, the program office will not be able to provide reliable timeline estimates or identify when problems or changes may occur and their effect on downstream work. Moreover, while the contractor conducted a schedule risk analysis on the schedule, that analysis was for select near-term milestones rather than the readiness of the ground system for the launch of JPSS-1 and it did not include the risks most likely to delay the project. A schedule risk analysis should be conducted through the finish milestone and should include risk data to determine activities that most often end up on the critical path.

Spacecraft Schedule

Of the ten best practices, the spacecraft schedule partially met eight best practices, and substantially met two best practices. There were strengths in the spacecraft schedule in that it was horizontally and vertically traceable; the contractor provided evidence of monthly progress updates to management, including status reporting of key milestones, handoffs, explanations of date changes, and an analysis of the critical and near critical paths; the contractor conducted a schedule risk analysis; and the schedule included baseline dates of activities for comparisons of actual performance to date.

However, there were also weaknesses in the spacecraft schedule. For example, the schedule had a low level of detail and included one-third of remaining activities with durations greater than 44 days, even after accounting for undefined and procurement-related activities. When establishing the durations of activities, they should be reasonably short and meaningful and allow for discrete progress measurement. Durations longer than 2 months do not facilitate objective measurement of accomplished effort and the milestone to detail activity ratio does not allow for effective progress measurement and reporting. As another example of a quality shortfall, the schedule was overly flexible with high float values that were not justified in schedule documentation. Specifically, 70 percent of remaining activities had about 5 business weeks of float, including 67 activities that had over 1,000 days of float, meaning that these activities could slip approximately 3.5 years without affecting the project's completion date. In order to establish reasonable total float, there should be documented justification for high float values in the schedule. Without this, it is unclear whether float values are high due to factors accepted by management and which are due to incomplete logic or other issues.

VIIRS Schedule

The VIIRS schedule partially met one best practice, substantially met seven best practices, and fully met two best practices. There were strengths in the VIIRS schedule in that the contractor established a clear process for integrating information between the schedule and resource management software, stated that durations were estimated by the people responsible for the activities based on work to be done, and justified in its schedule documentation activities with durations longer than 44 days. In addition, the contractor justified in schedule documentation the use of all date constraints, identified a valid driving path of activities for managing the program, and identified reasonable float values or justified them to the JPSS program office. Further, the contractor provided a schedule narrative accompanying each status update, which describes the status of key milestone dates (including the program finish date); explanations for changes in key dates; and a description of critical paths. However, there were also weaknesses in the VIIRS schedule. For example, the schedule had milestones that represented handoffs between contractor integrated product teams, but it did not include handoffs to the JPSS program office. In order to verify a schedule's horizontal traceability, handoffs should link products and outcomes associated with other sequenced activities. Without this, there could be different expectations between management and activity owners. As another example, the contractor conducted a schedule risk analysis with a good schedule network and obtained three different duration estimates from subject matter experts. However, the duration estimates did not reflect risks from the project's risk register and the analysis was focused only on activities on the critical path. This approach is flawed because activities that are not currently on the critical path could become critical as risks occur.

The inconsistency in quality among the three schedules has multiple causes. Program and contractor officials explained that certain weaknesses have been corrected with updated schedules. In other cases, the weaknesses lacked documented explanation in part because the JPSS program office did not require contractors to provide such documentation. Based on program schedule documentation, the schedule management and reporting requirements varied across contractors without documented justification for tailored approaches, which may partially explain the inconsistency in practices among the schedules. Since the reliability of an integrated schedule depends in part on the reliability of its subordinate schedules, schedule quality weaknesses in these schedules will transfer to an integrated master schedule derived from them. Consequently, the extent to which there are quality weaknesses in JPSS-1 support schedules further constrains the program's ability to monitor progress, manage key dependencies, and forecast completion dates. Until the program office addresses the scheduling shortfalls in its component schedules, the JPSS schedule will have lower quality and reduced reliability as a management tool for monitoring and forecasting satellite launch dates.

Program Has Confidence in the JPSS-1 Schedule, but Its Assumptions Do Not Reflect Weaknesses in the Underlying Data

According to our guidance on best practices in scheduling,[26] a schedule risk analysis uses statistical techniques to predict a level of confidence in meeting a program's completion date. This analysis focuses on key risks and how they affect the schedule's activities. The analysis does not focus solely on the critical path because, with risk considered, any activity may potentially affect the program's completion date. By relying on statistical simulations to randomly vary activity durations according to the probability of occurrence for certain durations and risks, the analysis seeks to develop a probability distribution of possible completion dates that reflect the program plan and enable an organization to match a date to its degree of risk tolerance.

The JPSS program office has conducted a schedule risk analysis on the JPSS-1 mission schedule (and launch date) through NASA's joint cost and schedule confidence level (JCL) process.[27] The JCL implemented by the JPSS program office represents a best practice in schedule management for establishing a credible schedule and reflects a robust schedule risk analysis conducted on key JPSS-1 schedule components. For example, the analysis assessed the impacts of key risks from the risk register and how multiple duration estimates for activities, based on documented uncertainty distributions, could affect the schedule. Based on the results of the JCL, the program office reports that its level of confidence in the JPSS-1 schedule is 70 percent and that it has sufficient schedule reserve to maintain a launch date of no later than March 2017.

However, the program office's level of confidence in the JPSS-1 schedule may be overly optimistic for two key reasons. First, the model that the program office used was based on flight project activities rather than an integrated schedule consisting of flight, ground, program office, and other activities relevant to the development and launch of JPSS-1. As a result, the JPSS program office's confidence level projections do not factor in the ongoing scheduling issues that are impacting the ground project. Had those issues been considered, the JPSS-1 confidence level would have been lower. Second, there are concerns regarding the spacecraft schedule's quality as discussed in the previous section. Factoring in these concerns, the confidence of the JPSS-1 satellite's schedule and projected launch date would be lower. We have previously reported that when using the JCL, NASA projects did not always include relevant cost and risk inputs.[28] While program officials noted that they included key ground system risks in their calculations, they did not include ground system scope in the JCL because it was too difficult to allocate ground system components to individual missions. Moreover, officials stated that they do not plan to include ground project or program office activities in future JCL updates. While it may have been difficult to include ground system scope in the JCL, without this, the program's schedule risk analysis and JCL do not reflect the full amount of work to be performed leading to JPSS-1 launch. Until the program office conducts a schedule risk analysis on an integrated schedule that includes the entire scope of effort and addresses quality shortfalls of relevant component schedules, it will have less assurance of meeting the planned March 2017 launch date for JPSS-1.

NOAA HAS ANALYZED ALTERNATIVES FOR ADDRESSING GAPS IN SATELLITE DATA, BUT LACKS A COMPREHENSIVE CONTINGENCY PLAN

While NOAA has identified multiple ways to help mitigate expected gaps in polar satellite data, it has not yet developed and implemented a comprehensive contingency plan. In October 2012, NOAA established a plan to address the impact of potential gaps in polar afternoon satellite data and contracted for a technical assessment that generated additional alternatives for the agency to consider. However, NOAA's mitigation plan has shortfalls when compared to government and industry best practices. Moreover, NOAA intends to update its plan by fall 2013 by integrating alternatives generated from the contractor's technical assessment. Until NOAA establishes a comprehensive contingency plan that addresses key shortfalls, it may not be positioned to effectively mitigate anticipated gaps in polar satellite coverage.

NOAA Identified Multiple Ways to Mitigate Polar Satellite Data Gaps

Polar satellites are essential to NOAA's mission to understand and predict changes in climate, weather, oceans, and coasts. Satellite data gaps in the morning or afternoon polar orbits would lead to less accurate and timely weather forecasting; as a result, advanced warning of extreme events would be affected. In June 2012, we reported that while NOAA officials communicated publicly and often about the risk of a polar satellite data gap, the agency had not established plans to mitigate the gap.[29] We recommended that NOAA establish mitigation plans for pending satellite gaps in the afternoon orbit as well as potential gaps in the early morning and midmorning orbits and NOAA agreed with the report's recommendation.

Table 9. Summary of NOAA's Polar Satellite Gap Mitigation Plan

Key assumptions	Alternatives	Key actions	Implementation status
There would be a polar afternoon gap of 14 to 18 months between March 2016 and October 2017 (the date that JPSS-1 is to become operational) Mission critical data from S-NPP's ATMS, CrIS, and VIIRS instruments would be lost DOD and European satellites would continue providing data in the early morning and midmorning orbits, respectively	Use similar data from available sources, such as existing DOD, NOAA, and NASA polar satellites Improve NOAA data assimilation Rely on foreign data, including radio occultation data or future polar satellites from other nations such as Russia or China Use non-satellite sources, such as aircraft observations Use commercial solutions (although none were identified)	Technical: • Conduct data denial experiments eliminating afternoon polar-orbiting sounder data from forecast models • Calculate, obtain, and distribute the estimated end-of-life of all sounder and imagery satellite assets Programmatic: • Monitor and report monthly on the health of instruments on existing polar satellites • Augment NOAA research and development computing capability as soon as possible to run data impact experiments	Actions were not implemented or funded.
		Management: • Commit to augmenting NOAA operational computing capability • Maintain international relationships that can result in partnerships for satellite data	

Source: GAO analysis of NOAA data.

In October 2012, NOAA established a mitigation plan to address the impact of potential gaps in polar afternoon satellite data. This plan identifies alternatives for mitigating the risk of a 14- to 18-month gap in the afternoon orbit beginning in March 2016, between the current polar satellite and the JPSS-1 satellite. Key alternatives include utilizing different satellites as data sources and improving data assimilation in models. The plan also lists technical, programmatic, and management actions needed to implement these options. Table 9 provides an overview of NOAA's polar satellite gap mitigation plan.

However, NOAA did not implement the actions identified in its mitigation plan and decided to identify additional alternatives. In October 2012, at the direction of the Under Secretary of Commerce for Oceans and Atmosphere (who is also the NOAA Administrator), NOAA contracted for a detailed technical assessment of alternatives to mitigate the degradation of products caused by a gap in satellite data in the afternoon polar orbit. This assessment solicited input from experts within and outside of NOAA and resulted in the following alternatives:

- rely on DOD's DMSP satellite;
- expand the use of radio occultation data, including funding the ground segment for a follow-on United States/Taiwan radio occultation mission;
- use atmospheric motion vectors (observed wind data);
- utilize future geostationary advanced imagery data;
- expand the use of aircraft observations;
- expand the use of targeted observations for high-impact events;
- implement a 4-dimensional hybrid data assimilation system (by adding a time dimension) ;
- improve data assimilation of cloud-impacted radiances;
- implement blends of global models, such as European and Canadian models;
- accelerate global model research to operations;
- sustain the use of high-latitude direct readout imagery; and
- rely on China's future Feng Yun-3 satellite.

Moving forward, NOAA officials stated that they are currently considering the additional alternatives and that the agency intends to integrate a final set of alternatives into its existing mitigation plan by the fall of 2013.

NOAA Does Not Yet Have a Comprehensive Contingency Plan

Government and industry best practices call for the development of contingency plans to maintain an organization's essential functions in the case of an adverse event.[30] As a complement to risk mitigation, contingency planning includes strategies that attempt to reduce or control the impact of risks should they occur. These practices identified by, for example, the National Institute of Standards and Technology and the Software Engineering Institute, include key elements such as defining failure scenarios, identifying and selecting strategies to address failure scenarios, developing procedures and actions to implement the selected strategies, testing the plans, and involving affected stakeholders. These elements can be grouped into categories, including (1) identifying failure scenarios and impacts, (2) developing contingency plans, and (3) validating and implementing contingency plans (see table 10).

Table 10. Guidelines for Developing a Sound Contingency Plan

Category	Description
Identifying failure scenarios and impacts	This category includes activities such as defining failure scenarios; conducting impact analyses that show the impact of failure scenarios; defining minimum acceptable levels of outputs and recovery time objectives; and establishing resumption priorities.
Developing contingency plans	This category includes activities such as identifying alternative solutions to address failure scenarios; selecting contingency strategies from among alternatives based on costs, benefits, and impacts; defining actions, roles and responsibilities, triggers, and timelines for implementing contingency plans; developing "zero-day" procedures; ensuring that steps reflect priorities for resumption of products and recovery objectives; and obtaining review and approval of the contingency plan from designated officials.
Validating and implementing contingency plans	This category includes activities such as identifying steps for testing contingency plans and conducting training exercises; preparing for and executing tests; validating test results for consistency against minimum performance levels; executing applicable actions for implementation of contingency strategies; communicating and coordinating with stakeholders to ensure that contingency strategies remain optimal foreducing potential impacts; and updating and maintaining contingency plans as warranted.

Source: GAO analysis of guidance documents from the National Institute of Standards and Technology, Software Engineering Institute, and GAO.

Table 11. Assessment of NOAA's Gap Mitigation Plan for its Polar Environmental Satellites

Contingency planning category	GAO assessment	Description
Identifying failure scenarios and impacts	Partially met	• The plan identifies key scenarios, such as an earlier than expected loss of data from the S-NPP satellite, a slip in the JPSS-1 launch date, a failure of JPSS-1 on launch, and a longer than expected calibration and validation period for JPSS-1. • The plan includes analyses of the impact to users from losing key weather products from ATMS, CrIS, and VIIRS. The plan also identifies minimum performance outputs for key weather data and reflects the top priorities identified in JPSS program requirements.
		• However, the plan does not address other scenarios, including the possibility of a loss of data from Department of Defense and European partner satellites in morning orbits or a partner mission in the afternoon orbit. • Further, the plan does not include recovery time objectives for key data products.
Developing contingency plans	Partially met	• The plan describes the impact of potential gaps in polar afternoon satellite data, identifies alternative strategies for mitigating the gap, and lists technical, programmatic, and management actions needed to implement gap mitigation strategies. • However, the plan has not yet been integrated with the other alternatives that were subsequently identified. • NOAA has not yet assessed its alternative strategies based on costs, benefits, and potential impacts. • The plan does not identify options for preventing gaps from occurring. • The plan does not identify opportunities for accelerating the calibration and validation phase—the time between launch and availability of operational products—on JPSS-1. • The plan does not identify specific actions for executing two of the five alternatives; identify roles and responsibilities for three alternatives; identify timelines for any of the alternatives; or identify triggers to signal when steps should be taken on any of the alternatives.
Validating and implementing contingency plans	Not met	• NOAA has not yet initiated efforts to validate or implement its gap mitigation plan.

Source: GAO analysis of NOAA data.

By documenting its mitigation plan and conducting a study on additional alternatives, NOAA has taken positive steps towards establishing a contingency plan for handling the potential impact of satellite data gaps in the afternoon polar orbit. However, NOAA does not yet have a comprehensive contingency plan because it has not yet selected the strategies to be implemented, or established procedures and actions to implement the selected strategies. In addition, there are shortfalls in the agency's current plans as compared to government and industry best practices, such as not always identifying specific actions with defined roles and responsibilities, timelines, and triggers. Moreover, multiple steps remain in testing, validating, and implementing the contingency plan. The following table provides an assessment of the extent to which NOAA's mitigation plan met contingency planning practices in three general categories.

NOAA officials stated that the agency is continuing to work on refinements to its gap mitigation plan, and that they anticipate issuing an updated plan in fall 2013 that will reflect additional alternatives. While NOAA expects to update its plan, the agency does not yet have a schedule for adding key elements—such as specific actions, roles and responsibilities, timelines, and triggers—for each alternative. Until NOAA establishes a comprehensive contingency plan that integrates its strategies and addresses the elements identified above to improve its plans, it may not be sufficiently prepared to mitigate potential gaps in polar satellite coverage.

CONCLUSION

While NOAA has made noteworthy progress over the past year in utilizing S-NPP data in weather forecasts and developing instrument and spacecraft components of the JPSS-1 satellite, the agency is facing challenges in its efforts to ensure sustained satellite observations. Specifically, NOAA does not expect to validate key S-NPP products until September 2014—nearly 3 years after the satellite's launch. Also, the agency does not track the usage of its satellite products or obtain feedback on them, which limits the program's ability to ensure that satellite products are useful. Further, the program experienced scheduling problems on its ground systems, which led to a delay in planned system upgrades. Until NOAA establishes a way to track which agencies are using its products and to obtain feedback on those products, the program office may continue to experience delays in delivering actionable S-NPP data to users.

Almost 3 years after the JPSS program was established, it lacks a complete integrated master schedule. While program officials recently established a preliminary integrated master schedule, the schedule lacks proper linkage among dependent activities, which limits its ability to calculate dates and predict changes in the future. Further, the quality of component schedules varied for certain practices. These issues raise questions about the program's 70 percent joint cost and schedule confidence level in the JPSS-1 launch date. Until the program office develops a complete integrated schedule and addresses weaknesses in component schedules, it will lack the information needed to effectively monitor development progress and ensure the planned JPSS-1 launch date.

NOAA has taken steps to mitigate an anticipated gap in polar afternoon satellite data, but its efforts are incomplete. Specifically, the agency has not yet established a comprehensive contingency plan that identifies specific actions with defined roles and responsibilities, timelines, and triggers for contingency strategies. Moreover, the agency's recent assessment of a larger set of alternatives has not yet been integrated with its mitigation plans. As a result, the agency faces important decisions as to whether and how the various alternatives should be carried out. While NOAA plans to add alternatives to its mitigation plan by fall 2013, it does not yet have plans to add the other key components. Until NOAA establishes a comprehensive contingency plan that addresses these shortfalls, its plan for mitigating potential gaps in the polar orbit may not be effective in avoiding significant impacts to NOAA's weather mission.

RECOMMENDATIONS FOR EXECUTIVE ACTION

Given the importance of having reliable schedules for managing JPSS satellite launch dates and the significance of polar-orbiting satellite data to weather forecasts, we recommend that the Secretary of Commerce direct the Administrator of NOAA to

- track the extent to which key groups of satellite data users are using S-NPP and JPSS products, and obtain feedback on these products;
- establish a complete JPSS program integrated master schedule that includes a logically linked sequence of activities;
- address the shortfalls in the ground system and spacecraft component schedules outlined in this report;

- after completing the integrated master schedule and addressing shortfalls in component schedules, update the joint cost and schedule confidence level for JPSS-1, if warranted and justified;
- establish a comprehensive contingency plan for potential satellite data gaps in the polar orbit that is consistent with contingency planning best practices identified in this report. The plan should include, for example, specific contingency actions with defined roles and responsibilities, timelines, and triggers; analysis of the impact of lost data from the morning orbits; and identification of opportunities to accelerate the calibration and validation phase of JPSS-1.

AGENCY COMMENTS

We sought comments on a draft of our report from the Department of Commerce and NASA. We received written comments from Commerce transmitting NOAA's comments. NOAA concurred with all five of our recommendations and identified steps that it is taking to implement them. It also provided technical comments, which we have incorporated into our report, as appropriate. NASA did not provide comments on the report's findings or recommendations, but noted that it would provide any input it might have to NOAA for inclusion in NOAA's comments.

David A. Powner
Director,
Information Technology Management Issues

APPENDIX I: OBJECTIVES, SCOPE, AND METHODOLOGY

Our objectives were to (1) evaluate the National Oceanic and Atmospheric Administration's (NOAA) progress in meeting the Joint Polar Satellite System (JPSS) program's objectives of sustaining the continuity of NOAA's polar-orbiting satellite system through the Suomi National Polar-orbiting Partnership (S-NPP) and JPSS satellites, (2) evaluate the quality of the JPSS program schedule, and (3) assess NOAA's plans to address potential gaps in polar satellite data.

To evaluate NOAA's progress in meeting JPSS program objectives, we assessed (1) the status of activities supporting the operational S-NPP satellite, (2) progress on efforts to develop the JPSS-1 satellite, and (3) recent changes in JPSS program scope. A more detailed description of our activities in each of these areas follows.

- **S-NPP progress:** We reviewed monthly program reports to identify the status of key upgrades to the ground system supporting S-NPP and the efforts to transition operational control of the satellite to NOAA. In addition, we compared the program's current estimated completion dates for S-NPP products to original program estimates for when the products would be available for operational use. We compared program office information on the extent to which S-NPP products were being used to best practices in evaluating the use of completed products. We also interviewed program officials about algorithm maturity and the extent to which users are using S-NPP products.
- **JPSS-1 progress:** We analyzed plans and reports on system development efforts for the JPSS-1 satellite. Specifically, we reviewed the JPSS-1 mission preliminary design review package to assess completion of work on the instruments, spacecraft, and ground system as well as cost, schedule, and technical performance for the JPSS-1 satellite. We also examined JPSS program office monthly status reports on system development progress to identify variances and corrective actions being taken to address the most critical issues and risks to the program. We interviewed JPSS program officials to discuss system development status. We assessed the reliability of reported milestone dates for top-level milestones by examining multiple project status reports at different points in time for consistent reporting of dates or explanations of any changes and compared reported dates to source schedule data. We determined that the milestone data were sufficiently reliable for our reporting purposes.
- **Changes in JPSS program scope:** We compared the program's requirements as of September 2011 to the program's updated plans and requirements as of May 2013 to identify key changes and to assess whether changes in capabilities have impacted program goals and objectives. We interviewed program officials about changes in the JPSS program's scope. We assessed the reliability of the program's estimated savings from program scope changes by comparing them to

program documentation on prior and current cost estimates and found that the estimates were sufficient for our purposes.

To evaluate the quality of NOAA's program schedule, we used an exposure draft of GAO's Schedule Assessment Guide[1] to assess schedule management practices and characteristics of selected contractor schedules. We selected and analyzed three component contractor schedules—the ground system, the spacecraft, and the Visible/Infrared Imager/Radiometer Suite instrument—because these schedules represented the critical path for flight and the entire ground system development schedule that was either already or likely to be driving the JPSS-1 satellite launch date. We also analyzed schedule metrics as a part of that analysis to highlight potential areas of strengths and weakness in, among other things, schedule logic, use of resources, task duration, float, and task completion. In order to assess each schedule against the ten best practices, we traced and verified underlying support and determined whether the program office or contractor provided a small portion, about half, a large portion, or complete evidence that satisfied the criterion and assigned a score depicting that the practices were met, minimally met, partially met, substantially met, or fully met. By examining the schedules against our guidance, we conducted a reliability assessment on each of the schedules and incorporated our findings on reliability limitations in the analysis of each component schedule. We reviewed documentation on a schedule risk assessment the JPSS program office conducted on JPSS-1 flight project schedules to identify assumptions and results of its analysis and to assess the reliability of the reported JPSS joint cost and schedule confidence level. We interviewed government and contractor officials to discuss reasons for observed shortfalls in schedule management practices. We determined that the schedules were sufficiently reliable for our reporting purposes and our report notes the instances where reliability concerns affect the quality of the schedules as well as the program's schedule risk assessment.

To assess plans to address potential gaps in polar satellite data, we reviewed NOAA's October 2012 polar satellite gap mitigation plan and a subsequent technical assessment as well as NOAA's plans for implementing recommendations from the assessment. We compared elements of the plan and assessment against best practices developed from leading government and industry sources such as the National Institute of Standards and Technology, the Software Engineering Institute's Capability Maturity Model® Integration, and our prior report. Based on that analysis, we identified shortfalls in NOAA's current plans as well as key remaining activities for the agency to accomplish. We interviewed NOAA headquarters staff and JPSS program officials about the technical assessment and their plans.

We performed our work at NASA and NOAA offices in the Washington, D.C. area. We conducted this performance audit from October 2012 through September 2013 in accordance with generally accepted government auditing standards. Those standards require that we plan and perform the audit to obtain sufficient, appropriate evidence to provide a reasonable basis for our findings and conclusions based on our audit objectives. We believe that the evidence obtained provides a reasonable basis for our findings and conclusions based on our audit objectives.

Appendix II: NOAA Plans to Transfer Selected JPSS Program Components to the Polar Free Flyer Program

In order to reduce Joint Polar Satellite System (JPSS) program costs and increase the program's focus on its weather mission, the National Oceanic and Atmospheric Administration (NOAA) plans to transfer key program components to a separate program, called the Polar Free Flyer program. After establishing JPSS in 2010, NOAA committed to developing three units of the Total and Spectral Solar Irradiance Sensor (TSIS) and to finding a spacecraft and launch accommodation for three instruments that would not be on the JPSS satellite: TSIS, the Advanced Data Collection System (A-DCS), and the Search and Rescue Satellite-Aided Tracking (SARSAT) system. As of June 2012, the JPSS program planned to launch two stand-alone satellites (called free flyers) to accommodate two suites of these instruments. However, NOAA recently made several decisions that affect these commitments, and expects to finalize these plans by the end of September 2013:

- NOAA plans to transfer responsibility for developing TSIS and accommodating the launch of the three instruments out of the JPSS program and into a newly established Polar Free Flyer program. According to JPSS program officials, a transition plan for the new program is under review and selected staff positions have been filled.
- The Polar Free Flyer program will deliver a single free flyer mission instead of the two missions planned under the JPSS program.

- NOAA will transfer the responsibility for developing the second TSIS instrument to the National Aeronautics and Space Administration (NASA), rely on an Air Force Global Positioning System mission to continue SARSAT coverage, and find a launch vehicle to accommodate an additional A-DCS instrument.
- NOAA plans to use the JPSS ground system to support the Polar Free Flyer Program.

The JPSS program plans to award a contract in fiscal year 2014 for a spacecraft that is to accommodate the TSIS, A-DCS, and SARSAT instruments. The three instruments are in development and testing, and are expected to be delivered to the satellite by 2015. The planned launch readiness date for the free-flyer mission was originally July 2016, but that date may change pending the outcome of the spacecraft contract award. Also, the program is looking to share a launch vehicle with some other mission to reduce launch costs. However, the program office is not aware of any ride-sharing opportunities that could accommodate the mission's planned launch readiness date.

APPENDIX III: ASSESSMENT OF JPSS COMPONENT SCHEDULES IMPLEMENTATION OF BEST PRACTICES IN SCHEDULING

The following tables identify detailed assessments of the extent to which three component schedules supporting the JPSS-1 schedule met the ten best practices and four characteristics of a high-quality, reliable schedule. Table 12 provides an assessment of the ground system contractor's schedule, which integrates activities from seven components of the ground system; table 13 provides an assessment of the spacecraft contractor's detailed schedule; and table 14 provides an assessment of the VIIRS contractor's detailed schedule.

The following information describes the key that we used in tables 12 through 14 to convey the results of our assessment of the schedules' consistency with an exposure draft of GAO best practices for schedule management.[1]

● Met: The program office or contractor provided complete evidence that satisfies the entire criterion.

◕ Substantially met: The program office or contractor provided evidence that satisfies a large portion of the criterion.

◑ Partially met: The program office or contractor provided evidence that satisfies about half of the criterion.

◔ Minimally met: The program office or contractor provided evidence that satisfies a small portion of the criterion.

○ Not met: The program office or contractor provided no evidence that satisfies any of the criterion.

Table 12. Detailed Assessment of Ground System Schedule Quality

Schedule characteristic or best practice	GAO assessment	Examples of strengths and weaknesses
Comprehensive		
Capturing all activities	◕	The schedule largely reflects the statement of work. However, the schedule only partially reflects the work breakdown structure and includes 40 activities that are marked as both summary activities and milestones.
Assigning resources to all activities	◕	The contractor has established a clear process for integrating information between the schedule and the resource management software. However, resource leveling has been performed outside of the schedule, which limits the effectiveness of the process.
Establishing the duration of all activities	◕	According to the contractor, durations were estimated by the people responsible for the activities based on work to be done. Additionally, calendars were used to specify valid working times for all activities.
		However, over 35 percent of the activities in the schedule were of long duration, and only half of these were justified in schedule documentation.

Schedule characteristic or best practice	GAO assessment	Examples of strengths and weaknesses
Well-constructed		
Sequencing all activities	◕	A majority of the activities in the schedule had dependencies, and the schedule's relationships were largely finish-to-start. However, program officials did not justify in schedule documentation the small number of activities with missing dependencies, date constraints, and lags.
Confirming that the critical path is valid	◔	The critical path and driving path are not fully valid because they are not free of long activities, constraints, and lags. Moreover, considering the schedule as a whole, the schedule software may not be calculating the true critical path of the project because the use of more than 800 constraints. These may result in float values that present an unrealistic view of the critical path.
Ensuring reasonable total float	◑	According to contractor officials, float values have been assessed as part of regularly scheduled health checks and they have determined that for certain cases float values are necessarily high. However, not all float values calculated by the schedule are reasonable and many values do not accurately reflect true schedule flexibility. Additionally, the JPSS program office did not provide a documented assessment of total float values that appear to be excessive to show that the team agrees with the logic and that the float is consistent with the plan.
Credible		
Verifying that the schedule can be traced horizontally and vertically	◑	The schedule is vertically traceable in all but one of the milestones that we reviewed, meaning that it allows activity owners to trace activities to higher-level milestones with intermediate and summary schedules.

Table 12. (Continued)

Schedule characteristic or best practice	GAO assessment	Examples of strengths and weaknesses
		However, the schedule is not fully horizontally traceable—that is, although the schedule includes giver/receiver milestones that are defined in the schedule documentation, the schedule was not always affected by activities whose durations were extended by hundreds of days.
Conducting a schedule risk analysis	◔	The contractor conducted a schedule risk analysis with a schedule network that partially meets the characteristics associated with a good schedule network, as well three point duration estimates that were captured from control account managers. However, the analysis was conducted for select near-term milestones—not to the readiness of the ground system for the launch of JPSS-1. Additionally, the analysis did not include risks most likely to delay the project, the paths or activities that are most likely to delay the project, and the activities that most often ended up on the critical path.
Controlled		
Updating the schedule using actual progress and logic	◕	Responsibility for changing the schedule has been assigned to someone who has the proper training and experience in critical path method scheduling and the schedule is free of clearly erroneous progress information.
		However, although the contractor provides a monthly program management briefing that addresses the status of external giver/receiver activities, it does not address the status of key milestone dates, changes in network logic, or critical paths.

Schedule characteristic or best practice	GAO assessment	Examples of strengths and weaknesses
Maintaining a baseline schedule	◕	A baseline schedule exists and is compared to the current schedule to track variances from the plan. According to contractor officials, a formal change control process is used to make changes to the baseline. However, the contractor's rolling wave reports do not satisfy all elements of a baseline schedule document. A baseline schedule document is a single document that describes, among other things, the organization of the IMS; the logic of the network; the basic approach to managing resources; the schedule's unique features; and justification for lags, date constraints, and long activity durations.

Source: GAO analysis of JPSS program office and contractor schedule data.

Table 13. Detailed Assessment of Spacecraft Schedule Quality

Schedule characteristic or best practice	**GAO assessment**	**Examples of strengths and weaknesses**
Comprehensive		
Capturing all activities	◑	The schedule reflects the work necessary to build the spacecraft, and schedule activities are mapped to the contract data requirements list and contractor work breakdown structure numbers. The schedule contains a low level of detail, which reflects the contractor's role as integrator for multiple vendors in a fixed-price environment. However, with a nearly 1:1 ratio of detail activities to milestones, the schedule would benefit from increased detail into work activities.
Assigning resources to all activities	◕	The contractor has established a clear process for integrating information between the schedule and the resource management software. However, resource leveling has been performed outside of the schedule, which limits the effectiveness of the process.

Table 13. (Continued)

Schedule characteristic or best practice	GAO assessment	Examples of strengths and weaknesses
Establishing the duration of all activities	◑	The contractor has experience in developing spacecraft similar to JPSS-1, including S-NPP. Contractor officials stated that they obtained duration estimates for activities from engineers that were responsible for them while other engineers conducted peer reviews on those estimates. However, durations in general appear too long to facilitate objective measurement of accomplished effort. Even accounting for procurement-related activities and level-of-effort type recurring meeting activities, one-third of all remaining activities are longer than 2 business months.
Well-constructed		
Sequencing all activities	◑	The schedule was partially logically sequenced. Approximately 20 percent of all remaining activities and milestones were missing predecessor links, successor links or both. Officials stated that many of these activities were related to contract data requirements list deliveries and internal or external handoffs (called givers/receivers). We found other areas of questionable sequencing logic. For instance, there are about 10percent of remaining activities in the schedule that have lags and leads, including some instances of leads with start-to-finish logic—a particularly abnormal logical relationship. We also found date constraints pervasive throughout the schedule: 140 activities have soft constraints and 17 have hard constraints.
		Hard constraints are useful for calculating the amount of float available in the schedule and, therefore, the realism of the required project finish date and available resources during schedule development. However, they may be abused if they force activities to occur on specific dates that are determined off-line without much regard for the realism of the assumptions necessary to achieve them.

Schedule characteristic or best practice	GAO assessment	Examples of strengths and weaknesses
Confirming that the critical path is valid	◑	The schedule defines activities with zero total float as critical. However, partly because of logic issues, the critical path as calculated by the scheduling software was convoluted and most likely unreliable. The path includes lags, leads, long-duration activities, and activities with hard constraints, which by definition will appear as critical. Officials stated they agreed that software-calculated critical paths cannot be relied upon in a complex schedule, and said they report the longest (or driving path) to management. Ideally, the critical path and the longest path should be the same, but our analysis found the longest path to be somewhat different than the default critical path; it does not include several activities that appeared on the critical path because of their date constraints. In addition, the longest path also includes several near-term, nonprocurement-related activities with long durations, spanning between 84 and 365 days.
Ensuring reasonable total float	◑	Officials stated that the total float values calculated by the schedule accurately reflect true schedule flexibility. However, we found that the schedule appears overly flexible due to high amounts of total float. 70 percent of remaining activities and milestones have greater than 30 days (about 5 business weeks) of total float. This includes 67 activities (8 percent of remaining) with over 1,000 days of float, meaning these activities can slip more than 3.5 business years before impacting the planned finish date of the project.
		Without documented justification for high float values in the schedule, it is not clear which are explained by milestones without successors, which are due to schedule maintenance, and which are due to incomplete logic.

Table 13. (Continued)

Schedule characteristic or best practice	GAO assessment	Examples of strengths and weaknesses
Credible		
Verifying that the schedule can be traced horizontally and vertically	◑	The schedule is vertically traceable, with dates in the detail schedule mapping to higher-level management briefing charts. The schedule is generally horizontally traceable. The schedule clearly identifies givers and receivers and negative total float calculations respond appropriately when significant delays are introduced into the network. However, negative float is calculated because key milestones are constrained. While the negative float may be an accurate assessment of potential delay, management may not be aware of potential delays when constrained dates are reported in summary-level schedules.
Conducting a schedule risk analysis	◑	Officials stated that they follow an internal process to perform schedule risk analyses on the schedule. Officials also stated that three-point durations are applied to activities, correlation is accounted for, and a Monte Carlo analysis is run on the schedule to derive probabilities for forecasted dates. Although the contractor has no contractual requirement to share schedule risk analysis results with the JPSS program office, it provided a summary of its risk assessment report and instructions. However, this summary information did not include supporting details such as risk data inputs and data normalization techniques and the contractor did not incorporate correlation or perform the schedule risk analysis on a logically sound (well-constructed) schedule.
Controlled		
Updating the schedule using actual progress and logic	◕	Schedule progress is updated monthly and the schedule is delivered to the JPSS program office in accordance with contractual requirements. While a formal schedule narrative does not accompany the schedule delivery to the government, much of the narrative information—such as the status of key milestones and handoffs,

Schedule characteristic or best practice	GAO assessment	Examples of strengths and weaknesses
		explanations for changes in key dates, and an overview of critical and near-critical paths—is conveyed in monthly management meetings. However, 26 activities had start or finish dates in the past. Of these, 12 activities could be explained by obsolete scope of work. We also found 12 out-of-sequence activities, representing 13 percent of in-progress activities.
Maintaining a baseline schedule	◑	Contractor officials stated that they maintain schedule baseline information in the default baseline fields in the schedule and we found that baseline dates were set in the schedule. However, a schedule baseline document was not created for the schedule baseline. We found 104 activities in the schedule without baseline dates, 72 of which are complete or are planned to start by 2014. The majority of start variances appear reasonable, but we did find start variances ranging from -221 days (221 days ahead of schedule) to 237 days (237 days delayed). Despite the significant variances noted, it is commendable that the schedule includes baseline information that allows for analysis and monitoring of dates' variances.

Source: GAO analysis of JPSS program office and contractor schedule data.

Table 14. Detailed Assessment of VIIRS Schedule Quality

Schedule characteristic or best practice	GAO assessment	Examples of strengths and weaknesses
Comprehensive		
Capturing all activities	◕	The schedule largely reflects the work breakdown structure and statement of work.
		However, the schedule does not reflect work to be performed by a subcontractor and includes 10 activities that are marked as both summary activities and milestones.
Assigning resources to all activities	◕	The contractor has established a clear process for integrating information between the schedule and the resource management software.

Table 14. (Continued)

Schedule characteristic or best practice	GAO assessment	Examples of strengths and weaknesses
		However, resource leveling has been performed outside of the schedule, which limits the effectiveness of the process.
Establishing the duration of all activities	●	According to the contractor, durations were estimated by the people responsible for the activities based on work to be done, realistic assumptions about available resources, productivity, normal interferences and distractions, and reliance on others. Further, the contractor justified in its schedule documentation virtually all activities with durations longer than 44 days.
Well-constructed		
Sequencing all activities	◕	All but one activity in the schedule has at least one predecessor and one successor, and that activity was justified in the schedule documentation. Additionally, every schedule date constraint was justified in schedule documentation. However, the schedule has a very small number of activities with dangling logic. Further, although explanations were provided for most of the small number of lags, the explanations did not justify their use.
Confirming that the critical path is valid	◕	Program office and contractor officials use the driving path to manage the program, which is preferred because it represents the activities that are driving the sequence of start dates directly affecting the estimated finish date. However, the driving path and the critical path to key milestones should be the same, and they are not. Also, the critical path is not valid because it contains level of effort activities.
Ensuring reasonable total float	◕	The program office has defined reasonable float values, and the values associated with the schedule largely fit that definition. For those float values that were not reasonable, the program office provided a documented assessment of those values to show that the team agrees with the logic and that the float is consistent with the plan. However, the schedule has a small number of activities that have unrealistic float values.

Schedule characteristic or best practice	GAO assessment	Examples of strengths and weaknesses
Credible		
Verifying that the schedule can be traced horizontally and vertically	◕	The schedule is largely horizontally traceable. In particular, the schedule is affected by activities whose durations are extended by hundreds of days, and it includes giver/receiver milestones that represent handoffs between contractor integrated project teams. However, the schedule does not include all givers/receivers between the contractor and the program office. Additionally, the schedule is vertically traceable. Specifically, it allows activity owners to trace activities to higher-level milestones with intermediate and summary schedules.
Conducting a schedule risk analysis	◑	A schedule risk analysis was conducted with a good schedule network, and three point duration estimates that were captured from subject matter experts. However, the duration estimates did not reflect risks from the project's risk register and the analysis was focused on only the deterministic critical path and near-critical path.
Controlled		
Updating the schedule using actual progress and logic	●	Responsibility for changing or updating the schedule has been assigned to someone who has the proper training and experience in critical path method scheduling. Additionally, the schedule is free of clearly erroneous progress information.
		Further, the contractor provides a schedule narrative accompanying each status update, which describes the status of key milestone dates (including the program finish date); explanations for changes in key dates; and a description of the critical paths.
Maintaining a baseline schedule	◕	A baseline schedule exists and is compared to the current schedule to track variances. However, the contractor did not have a baseline schedule document. A baseline schedule document is a single document that describes, among other things, the organization of the IMS; the logic of the network; the basic approach to managing resources; the schedule's unique features; and justification for lags, date constraints, and long activity durations.

Source: GAO analysis of JPSS program office and contractor schedule data.

End Notes

[1] *GAO Schedule Assessment Guide: Best Practices for Project Schedules*, GAO-12-120G (exposure draft) (Washington, D.C.: May 30, 2012).

[2] See GAO, *Year 2000 Computing Crisis: Business Continuity and Contingency Planning*, GAO/AIMD-10.1.19 (Washington, D.C.: August 1998); National Institute of Standards and Technology, *Contingency Planning Guide for Federal Information Systems,* NIST 800-34 (May 2010); Software Engineering Institute, *CMMI® for Acquisition, Version 1.3* (Pittsburgh, Pa.: November 2010).

[3] NOAA provides command and control for both the POES and DMSP satellites after they are in orbit.

[4] The European Organisation for the Exploitation of Meteorological Satellites' MetOp program is a series of three polar-orbiting satellites dedicated to operational meteorology. MetOp satellites are planned to be flown sequentially over 14 years. The first of these satellites was launched in 2006, the second was launched in 2012, and the final satellite in the series is expected to launch in 2017.

[5] Presidential Decision Directive NSTC-2, May 5, 1994.

[6] Originally called the NPOESS Preparatory Project, in January 2012, the satellite's name was changed to the Suomi National Polar-orbiting Partnership satellite.

[7] GAO, *Polar-orbiting Operational Environmental Satellites: Technical Problems, Cost Increases, and Schedule Delays Trigger Need for Difficult Trade-off Decisions*, GAO-06-249T (Washington, D.C.: Nov. 16, 2005).

[8] Using S-NPP as an operational satellite means that the satellite's data will be used to provide climate and weather products.

[9] The announcement accompanied the release of the President's fiscal year 2011 budget request.

10 This figure does not include approximately $2.9 billion in sunk costs that DOD spent on NPOESS through fiscal year 2010.

[11] NOAA's $3.3 billion sunk costs included $2.9 billion through fiscal year 2010 and about $400 million in fiscal year 2011.

[12] See, for example, GAO, *Polar Satellites: Agencies Need to Address Potential Gaps in Weather and Climate Data Coverage*, GAO-11-945T (Washington, D.C.: Sept. 23, 2011); *Polar-orbiting Environmental Satellites: Agencies Must Act Quickly to Address Risks That Jeopardize the Continuity of Weather and Climate Data,* GAO-10-558 (Washington, D.C.: May 27, 2010); *Polar-orbiting Environmental Satellites: With Costs Increasing and Data Continuity at Risk, Improvements Needed in Tri-Agency Decision Making,* GAO-09-772T (Washington, D.C.: June 17, 2009); and *Polar-orbiting Environmental Satellites: With Costs Increasing and Data Continuity at Risk, Improvements Needed in Tri-Agency Decision Making*, GAO-09-564 (Washington, D.C.: June 17, 2009).

[13] GAO, *Polar-orbiting Operational Environmental Satellites: Changing Requirements, Technical Issues, and Looming Data Gaps Require Focused Attention*, GAO-12-604 (Washington, D.C.: June 15, 2012).

[14] GAO, *High-Risk Series: An Update*, GAO-13-283 (Washington, D.C.: February 2013).

[15] The European Centre for Medium Range Weather Forecasts is an independent, intergovernmental organization supported by 34 European nations, providing global medium-to-extended range forecasts.

[16] According to NOAA and NASA officials, the products go through a beta stage (in which products have been minimally validated, but are available to users so that they can begin working with the data); a provisional stage (in which products are not optimal, but are ready

for operational evaluation by users); and a validated stage (in which products are ready for operational use).

[17] Software Engineering Institute, *CMMI® for Acquisition, Version 1.3* (Pittsburgh, Pa.: November 2010) and GAO, *Information Technology Investment Management: A Framework for Assessing and Improving Process Maturity*, Version 1.1, GAO-04-394G (Washington, D.C.: March 2004).

[18] Executive Office of the President National Science and Technology Council, *National Strategy for Civil Earth Observations* (Washington, D.C.: April 2013).

[19] See GAO-12-120G (exposure draft).

[20] GAO-12-604.

[21] In January 2013, program officials revised this delay to 96 minutes to more precisely reflect the time it takes to send products from the ground system to the end users.

[22] GAO-12-120G (exposure draft).

[23] GAO-12-120G (exposure draft).

[24] The critical path is generally defined as the longest continuous sequence of activities in a schedule. As such, it defines the program's earliest completion date or minimum duration.

[25] These three component schedules represent the critical path for the flight project and the entire ground system schedule.

[26] GAO-12-120G (exposure draft).

[27] The JCL is a probabilistic analysis that includes, among other things, all cost and schedule elements, incorporates and quantifies potential risks, assesses the impacts of cost and schedule to date, and addresses available annual resources to arrive at development cost and schedule estimates associated with various confidence levels.

[28] GAO, *NASA: Assessments of Selected Large-Scale Projects*, GAO-12-207SP (Washington, D.C.: March 1, 2012).

[29] GAO-12-604.

[30] See GAO, *Year 2000 Computing Crisis: Business Continuity and Contingency Planning*, GAO/AIMD-10.1.19 (Washington, D.C.: August 1998); National Institute of Standards and Technology, *Contingency Planning Guide for Federal Information Systems,* NIST 800-34 (May 2010); Software Engineering Institute, *CMMI® for Acquisition, Version 1.3* (Pittsburgh, Pa.: November 2010).

End Note for Appendix I

[1] *GAO Schedule Assessment Guide: Best Practices for Project Schedules*, GAO-12-120G (exposure draft) (Washington, D.C.: May 30, 2012).

In: Weather Satellites
Editor: Sherril G. Ryder
ISBN: 978-1-62948-687-1

Chapter 3

STATEMENT OF MARY E. KICZA, NESDIS ASSISTANT ADMINISTRATOR, NATIONAL OCEANIC AND ATMOSPHERIC ADMINISTRATION. HEARING ON "DYSFUNCTION IN MANAGEMENT OF WEATHER AND CLIMATE SATELLITES"*

Good morning Chairmen Stewart and Broun, Ranking Members Bonamici and Maffei, and Members of the Committee. I am Mary Kicza, the Assistant Administrator of NOAA's National Environmental Satellite, Data, and Information Service (NESDIS). Thank you for the opportunity to join Mr. David Powner from the Government Accountability Office (GAO), and Mr. Marcus Watkins from the National Aeronautics and Space Administration (NASA) at today's hearing. While I will focus my remarks on NOAA's Joint Polar Satellite System (JPSS) and Geostationary Operational Environmental Satellite-R (GOES-R) Series Programs, I am ready to address questions on NOAA's entire satellite acquisition portfolio.

NOAA's mission to provide science, service, and stewardship to the Nation is fundamentally dependent on observations of our environment. These observations are the backbone of NOAA's predictive capabilities. NOAA must ensure operational weather, ocean, climate, and space weather data are

* This is an edited, reformatted and augmented version of a statement, presented September 19, 2013 before the House Committee on Science, Space, and Technology, Subcommittees on Environment and Oversight.

available seven days a week, 24 hours a day, to address our Nation's critical needs for timely and accurate forecasts and warnings of solar storms and severe weather, such as hurricanes, flash floods, tsunamis, winter storms, and wildfires. Of the data actually assimilated into NOAA's National Weather Service (NWS) numerical weather prediction models that are used to produce the longer term weather forecasts three days and beyond, over 95 percent comes from satellites, of which over 80 percent are from polar-orbiting satellites. These polar-orbiting satellites include NOAA's Polar-orbiting Operational Environmental Satellite (POES), Suomi National Polar-orbiting Partnership (Suomi NPP) satellite, and NASA Earth Observing Satellites (EOS) in the afternoon orbit, and the European Metop satellites which fly in the mid-morning orbit. GOES satellites, along with Doppler Radar, assist operational weather forecasters with current and short-term forecasting abilities (i.e., weather that is occurring now up to three days in the future) and severe weather warning forecasts.

The American public and the commercial sector expect that they can continue to rely on receiving accurate, reliable, and timely weather information from NWS. The growing private weather sector, which delivers specialized weather information, is another consumer reliant on receipt of these data. NOAA's satellites are an integral part of the observational infrastructure that supports these NWS and commercial sector forecasting capabilities.

My testimony today will focus on the progress that NOAA, with NASA as our acquisition agent, is achieving in developing the Nation's next generation geostationary and polar-orbiting satellite systems, the GOES-R Series and JPSS Programs. While these geostationary and polar-orbiting systems provide the backbone of data that drive the NWS numerical weather prediction models, NOAA augments its observational needs by leveraging data from research satellites (e.g., NASA EOS and Advanced Composition Explorer); by using data from Department of Defense satellites (e.g., Defense Meteorological Satellite Program); by purchasing data from the commercial sector (e.g., lightning data and space-based Synthetic Aperture Radar); by implementing international agreements to ingest data from partner organizations (e.g., Metop data, Constellation Observing System for Meteorology, Ionosphere and Climate (COSMIC) data); and by jointly procuring satellite systems through domestic (e.g., Deep Space Climate Observatory (DSCOVR) refurbishment) and international partnerships (e.g., Jason-3 acquisition).

Over the past five years, NOAA has made strategic and deliberate steps to return to an over 40 year partnership with NASA for acquisition of NOAA's operational satellites. This renewed partnership leverages the acquisition and developmental expertise of NASA to provide systems engineering and procurement support on a fully reimbursable basis. This decision continues a partnership that has supported an unprecedented series of successful launches of NOAA's geostationary and polar-orbiting operational satellites and enables in-depth technical management and systems engineering oversight of NOAA's satellite acquisition portfolio.

I am pleased to report this partnership continues to enable us to stabilize management structures, staffing, requirements, and program oversight. This has led to the completion of key program milestones, and gives us confidence that we will meet the cost, schedule, and performance milestones that lie ahead.

Our recurring challenge is being able to plan and execute our programs in the difficult budget environment we all face at this time. This requires the timely receipt of appropriated funds of the requested amounts, allowing these critical satellite systems to maintain their current development cadence, thereby ensuring the GOES-R Series and JPSS satellites are successfully launched and operating before the current legacy satellites cease to provide useful data. Without this, NOAA cannot guarantee that cost, schedule, and performance commitments can be met.

Over the past year, we were reviewed by David Powner's team, the Information Technology Team, at the GAO, the Department of Commerce Office of Inspector General (OIG), and independent review teams comprised of eminent aerospace experts. While the reviews looked at many specific areas of concern, all reviews urged us to remain focused on maintaining the continuity of our observational capability and being prepared to mitigate the impacts of a gap in coverage. The reviewers were mindful of the immediate and devastating impacts to the Nation that a gap in coverage would bring, starting with degraded forecast quality and skill, which could place lives and property at risk from severe weather events. We concur with the recommendations from these groups and are working to ensure the GOES-R Series and JPSS satellite development continues. The Committee has expressed concern that GAO added NOAA's satellite acquisition of the GOES-R Series and JPSS Programs to its Biennial High Risk List due to the risk of gaps in weather observations. The steps that NOAA has been taking over the years to implement the OIG and GAO recommendations, and will take to implement these most current recommendations will address many of the root causes that the GAO High Risk report emphasized, specifically, the potentials for gaps in coverage and the impact on weather forecasts and on the customers and users of NOAA's satellite data. Additionally, the 2012 Independent Review Team (IRT) led by Thomas Young also provided invaluable recommendations that informed NOAA's FY 2014 Budget request and underscored the importance of refocusing the JPSS Program on a weather mission.

PROGRESS ON THE GOES-R SERIES PROGRAM

The GOES-R Series Program is NOAA's next-generation geostationary environmental satellite constellation. Geostationary environmental satellites are our observational sentinels in space, providing constant watch for severe weather such as hurricanes, severe thunderstorms, flash floods, and wildland fires in the Western Hemisphere. With two geostationary satellites always in operation (GOES-East and GOES-West) and an on-orbit spare, we are able to track severe weather from the coast of Africa across to most of the Pacific basin. The GOES satellites complement *in situ* observational systems such as NOAA's Doppler Radar network, NOAA's Hurricane Hunters, and ocean buoys to provide NWS forecasters with near real-time data used to support operational weather forecasts.

GOES-R Series Program Content. The GOES-R Series Program content remains unchanged since the 2012 hearing, and progress is being made on the development of all the instruments which will provide continuity and needed enhancements of required weather and space weather data:

- Advanced Baseline Imager (ABI)
- Geostationary Lightning Mapper (GLM)
- Space Environmental In-Situ Suite (SEISS)
- Extreme Ultra Violet / X-Ray Irradiance Sensor (EXIS)
- Solar Ultra Violet Imager (SUVI)
- Magnetometer

GOES-R Series Program Progress. The GOES-R Series Program has made significant progress in its development, including:

- Delivery of the first GOES-R instrument, the Extreme Ultraviolet and X-ray Irradiance Sensor (EXIS) Flight Model 1 (FM1), which is ready to be integrated onto the GOES-R spacecraft.

- Completion of environmental testing of the Advanced Baseline Imager (ABI), the Solar Ultraviolet Imager (SUVI), and the Space Environment In-Situ Suite (SEISS). The ABI, SUVI, and SEISS will all be available for integration with the spacecraft by the end of 2013.
- Progress with the remaining GOES-R instruments, the Geostationary Lightning Mapper (GLM) and Magnetometer, which are on track for delivery in mid-2014.
- Significant progress on the development of the GOES-R spacecraft bus, which is on track to be available for instrument integration in 2014.
- Delivery of the first significant release of Core Ground System to the Government's Mission Operations Support Team located at the NOAA Satellite Operations Facility in Suitland, Maryland.
- Progress with the assembly and installation of GOES-R antenna structures at the Wallops Command and Data Acquisition Station in Wallops Island, Virginia and at the Remote Backup Unit (RBU) in Fairmont, West Virginia. These efforts are proceeding on schedule with four of six new antenna structures already complete.

GOES-R Series Program Cost. The GOES-R Series Program Baseline Report, dated February 8, 2013, reported to Congress that the lifecycle cost (LCC) for the GOES-R Series Program is $10.860 billion for four satellites. Maintaining the GOES-R Series Program at the LCC reflected in the Baseline Report is dependent on NOAA receiving appropriations as requested in the Administration's annual budget requests. Given sequestration and rescission impacts, the FY 2013 enacted appropriations bill did not provide the level of funding required for the GOES-R Series Program as outlined in the Baseline Report. The GOES-R Series Program is still refining the cost impact of the FY 2013 appropriations bill to the LCC.

GOES-R Series Schedule. The GOES-R Series Program recently completed an assessment of the impact on the schedule due to the $54 million reduction from the sequestration and rescission included in the enacted FY 2013 appropriations. The effects from these budget reductions continue to be assessed. The GOES-R Series Program continues, however, to work aggressively in order to have GOES-R and GOES-S ready as early as possible. We are confident that we can meet these adjusted launch date commitments for GOES-R and GOES-S given timely receipt of requested funding amounts moving forward.

2013 GAO Review of the GOES-R Series Program

NOAA was provided an opportunity to review the draft GAO recommendations and NOAA concurs with the four GAO recommendations for the GOES-R Series Program reflected in that document. We will review the final report and the recommendations contained therein and will work to address them.

These recommendations include direction to:

- include information and methodology for calculating the amount of reserve funding for each of the four satellites in the program in regular briefings to NOAA senior executives;
- improve sequencing of all activities to ensure there are adequate resources for the activities, and conduct a schedule risk analysis on the core ground systems and spacecraft schedules;
- improve communications with internal and external satellite data users on changes in GOES-R requirements;
- revise the satellite and ground system contingency plans to provide more information on the potential impact of a satellite failure, identifying alternative solutions for preventing a delay in GOES-R launch as well as timelines for implementing those solutions, and coordinating with key external stakeholders on contingency strategies.

With sufficient appropriations, as noted above, we will minimize the possibility of a gap of coverage in the geostationary orbit.

2013 OIG Review of the GOES-R Series Program

The OIG completed its most recent review of the GOES-R Series Program in April 2013. NOAA has developed action plans which it is implementing in response to these findings and recommendations.

Recommend that the NOAA Deputy Under Secretary for Operations:

1. Develop a comprehensive set of tradeoff approaches to mitigate launch delays and communicate approaches to stakeholders and users.

2. Keep stakeholders and users informed of tradeoffs made to meet the launch date.
3. Direct NESDIS to report periodically on the adequacy of program systems engineering integration and NASA systems engineering support.

Recommend that the NOAA Assistant Administrator for NESDIS ensure that NASA:

4. Effectively validates contractors' proposals and subsequent plans, to verify that technical designs meet readiness requirements per NASA standards.
5. Modifies contract award-fee structures to reduce award fee percentages and clearly articulates how scores should be adjusted based on the magnitude of cost overruns.
6. Adjusts future award fees to be more commensurate with contractor performance.

Recommend that the NOAA Deputy Under Secretary for Operations:

7. Direct the development of a policy for managing undefinitized contract actions to definitize change orders in the shortest practicable time.

We believe that implementing these recommendations will improve our overall management of the GOES-R Series Program.

2012 IRT Review

While the IRT provided an overall review of NOAA's satellite portfolio, the team stated that the GOES-R model of governance is most consistent with the principles for success, and that its program requirements were well defined for its weather and space weather mission.

There were other recommendations such as streamlining reporting and the decision making processes that have benefited the GOES-R Series Program. The IRT cautioned that the GOES-R gap required vigilance and management attention.

Progress on the JPSS Program

NOAA's Polar-orbiting Operational Environmental Satellite (POES) series provides full global coverage for a broad range of weather and environmental monitoring applications. Placed in the afternoon orbit, NOAA POES and NASA EOS satellites are crucial for NWS's three to seven day weather forecasts and environmental modeling efforts. The current POES satellite, NOAA19, was launched in early 2009 with a design life lasting until the second quarter of FY 2013. While it continues to operate, it is beginning to demonstrate some sensor deterioration. The Suomi NPP satellite, which was launched in October 2011 with a design life lasting until the first quarter of FY 2017 and is now operated by NOAA as a part of the JPSS Program, is providing operational data to the NWS weather forecast models. Both NOAA-19 and Suomi NPP are providing coverage for the afternoon orbit. The European Metop satellite constellation (Metop A and Metop B), which flies in the mid-morning orbit, is also robust and NOAA uses these data in its numerical weather prediction models.

JPSS Program Content. The JPSS Program, as described in the President's FY 2014 Budget request, consists of three satellites: Suomi NPP, JPSS-1 and JPSS-2, and the associated ground system.

The JPSS Program has been refocused to support a weather mission and will fly the following core instruments:

- Advanced Technology Microwave Sounder (ATMS)
- Cross-track Infrared Sounder (CrIS)
- Visible Infrared Imaging Radiometer Suite (VIIRS)
- Ozone Mapping and Profile Suite (OMPS)-Nadir

While JPSS-1 will fly the Clouds and the Earth's Radiant Energy System (CERES) instrument, responsibility for funding the development of the follow on instrument has been transferred to NASA. NASA will also assume responsibility for funding continuity of the second Total Solar and Spectral Irradiance Sensor (TSIS) and the OMPS-Limb instruments. The responsibility for providing spacecraft bus and rideshare for the first TSIS, Satellite-assisted Search and Rescue instrument (SARSAT), and the Advanced Data Collection System (A-DCS) has been transferred out of the JPSS portfolio to a new NOAA program entitled Polar Free Flyer.

JPSS Program Progress. Over the past year, the JPSS Program has achieved the following successes:

- Completed and passed the DOC Milestone 2/3 and Key Decision Point (KDP)-I reviews which documented the cost and schedule commitment for accomplishing the development and operation of JPSS.
- Submitted, as part of the President's FY 2014 Budget request, a JPSS Program focused on NOAA's weather mission, revising the total program LCC to $11.3 billion.
- Completed reconciliation with a DOC Independent Cost Estimate, which validated the revised program cost.
- Completed JPSS-1 Mission KDP-C in July 2013, which baselined its cost, performance, and schedule to launch in early 2017.
- Transitioned Suomi NPP Ground Operations from NASA to NOAA in February 2013.
- Completed assembly of all JPSS-1 instruments and advanced them to instrument level test phase.

JPSS Program Cost. Since the 2012 hearing before this Committee, the Administration has reduced the scope and content of the JPSS program to more narrowly focus on the weather mission. This reduction was achieved by transferring responsibility for non-weather instruments to NASA and other programs within NOAA, trimming content, and improving efficiency. This has resulted in a reduction of the JPSS Program LCC from $12.9 billion as reflected in the President's FY 2013 Budget request to the current LCC of $11.3 billion. This new cost commitment was affirmed by the recent KDP-I approval. That figure consists of the prior year costs of $2.5 billion legacy costs from the earlier NPOESS program and $2.6 billion appropriated from FY 2010 through FY 2013 to implement the transition and beginning of the JPSS program, and a commitment of $6.2 billion from FY 2014 to FY 2025. The latter total is required to operate the Suomi NPP satellite, finish building the JPSS-1 satellite, build the JPSS-2 satellite, continue development of the ground system, and perform the activities required to launch JPSS-1 and JPSS-2 and operate and sustain all three satellite missions in the JPSS portfolio, i.e., Suomi NPP, JPSS-1, and JPSS-2.

JPSS Schedule. The launch schedule for the JPSS-1 satellite in the second quarter of FY 2017 remains unchanged since the 2012 hearing. We have recently completed additional analyses which indicate that we are in a much better position to meet this schedule based on the progress made on ground and flight development and based on full funding provided in the FY 2014 House and Senate Marks. With respect to the JPSS-2 launch readiness date, with the refocusing of the JPSS Program to weather mission, NOAA has rephased out-year funding to bring forward the JPSS-2 launch to the first quarter of FY 2022.

2013 GAO Review of the JPSS Program

NOAA was provided an opportunity to review the draft GAO recommendations and NOAA concurs with the five GAO recommendations for the JPSS Program reflected in that document. We will review the final report and the recommendations contained therein and will work to address them.

The recommendations include direction to:

- track the extent to which groups of satellite data users are using Suomi NPP and JPSS products and obtain feedback on these products;
- establish a complete JPSS Program integrated master schedule that includes a logically linked sequence of activities;
- address the shortfalls in the ground system and spacecraft component schedules outlined in the report;
- update the joint cost and schedule confidence level for JPSS-1, if warranted and justified after completing the integrated master schedule and addressing shortfalls in component schedules; and
- establish a comprehensive contingency plan for potential satellite data gaps in the polar orbit that is consistent with the contingency planning best practices identified in this report. The plan should include, for example, specific contingency actions with defined roles and responsibilities, timelines, and triggers; analysis of the impact of lost data from the morning orbits, and identification of opportunities to accelerate the calibration and validation phases of JPSS-1.

Refocusing the JPSS Program to a weather mission and moving content to other programs has improved our confidence on meeting the second quarter FY 2017 launch readiness date for the JPSS-1 satellite, thereby minimizing the possibility of gaps in data coverage noted in the GAO's "High Risk" report.

While there is still a risk of a gap in coverage, recent analyses and assessments have increased our confidence that we will launch JPSS-1 in the second quarter of FY 2017. This, coupled with a rigorous management regime for the Suomi NPP satellite to preserve operating life, gives us confidence that if the satellite continues to perform as expected, we will significantly reduce risk of a gap of coverage in the afternoon orbit. However, sufficient funding is required to ensure that we maintain the current acquisition schedule.

2013 OIG Review of the JPSS Program

The OIG has not yet completed its 2013 review. The JPSS Program continues to provide full and open access to the employees and records to the OIG as they carry out their review. We remain ready to review their findings and recommendations.

2012 IRT Review

The IRT findings and recommendations were organized in five areas of concern:

1. Oversight and Decision Process
2. Governance
3. JPSS Gap
4. Programs, and
5. Budget

Significant changes have been implemented at the Department of Commerce, NOAA, and NASA in response to the IRT recommendations. Most notably, the JPSS program, as reflected in the President's FY 2014 Budget request, was modified to focus on NOAA's weather mission, with the JPSS-2 schedule for launch accelerated. Oversight, decision processes and governance have been both streamlined and clarified. The DOC, NOAA, and NASA teams are currently in the process of reviewing with the IRT our responses to their recommendations.

Looking to the Future

As we look to the future, NOAA is aware that the challenges we face require examining and reaffirming the core activities that the Government must provide and excel at, while looking at possible roles that partnership with the commercial sector and international community can provide. For example, this Committee has issued a challenge to NOAA to extend tornado warning lead times to one hour. As we work to address the fundamental warning paradigm shift this challenge would entail, we will require input from the entire NOAA enterprise. As a major data provider for weather forecasting, NESDIS is working with NWS and NOAA's Office of Oceanic and Atmospheric Research to assess the best means of approaching this challenge from scientific, technological, and human behavioral dimensions.

A first and crucial step towards meeting this challenge is to minimize any potential for gaps in coverage between the handoff from the current capabilities that NOAA's satellite constellation provide to the next generation GOES-R and JPSS Programs. In addition, NOAA remains focused on retaining our current forecasting skill and making incremental improvements towards providing as much advance warning as we can to protect lives and safeguard property from severe weather events.

The importance of NOAA's relationship with the aerospace industry is another integral aspect in meeting such challenges. Most of the instruments, all of the spacecraft buses, and much of the ground systems and data processing capabilities that will support the GOES-R Series and JPSS Programs are being developed by the aerospace industry under contract to either NASA or NOAA. Moreover, NOAA is considering commercial rideshare opportunities for the first TSIS instrument, satellite assisted search and rescue (SARSAT) and the Advanced Data Collection System (A-DCS) instruments.

On the international front, NOAA continues to support and strengthen international partnerships that help us to maintain continuity of geostationary and polar-orbiting satellite observational capability. To that end, I offer the following important updates:

- NOAA signed a new agreement with its European operational satellite counterpart, the European Organisation for the Exploitation of Meteorological Satellites (EUMETSAT) that provides a broad policy framework for all of its wide-ranging cooperative activities. Building on a 30-year relationship, this agreement commits both to continued cooperation in the areas of geostationary backup support, joint ocean altimetry satellite development and operation (Jason series), and our JPSS, whereby NOAA and EUMETSAT each operate a polar-orbiting satellite in one of the two key orbits required for weather forecasting (NOAA will operate JPSS in the "afternoon" orbit).

- NOAA is working with the Japan Aerospace Exploration Agency (JAXA) to receive and process data from its first Global Change Observation Mission-Water (GCOM-W1) satellite, which carries the Advanced Microwave Scanning Radiometer (AMSR-2) instrument. These data help NOAA conduct assessments of precipitation, water vapor amounts, wind velocity above the ocean, sea surface temperature, water levels on land areas, and snow depths.
- Partnerships with the Canadian and French Space Agencies have been reaffirmed, with these agencies providing the satellite-assisted search and rescue (SARSAT) and the Advanced Data Collection System (A-DCS) instruments planned for launch by the NOAA Polar Free Flyer Program.
- NOAA has reaffirmed partnerships with our European and Japanese space partners for backup support in the event of a catastrophic loss of NOAA GOES or POES/Suomi NPP satellites.

With funds provided by the Public Law 113-2, "*Disaster Relief Appropriations Act of 2013,*" NOAA is implementing a number of strategic actions designed to make its weather forecasting enterprise more robust in the face of the possibility of a gap in polar-orbiting weather data. These activities seek to make better use of existing data, take advantage of new data sources planned in the future, improve operational high performance computing capacity, and improve the assimilation of data into weather prediction models, including hurricane models. The goal is to minimize the impact of a gap in coverage should it become a reality. While none of these activities, individually or collectively, can totally replace a lack of JPSS data, they represent the positive actions NOAA can take to mitigate the loss of these data. Should a data gap not occur, these investments will nonetheless improve NOAA's ability to use existing data, thus improving weather forecasts. These actions are being taken in addition to the steps NOAA is taking to ensure that JPSS and GOES-R Series satellite development continue as planned.

Meeting the customers' needs remains the most important part of this acquisition process. As the GOES-R Series Program enters the phase where the first satellite will be launched, it has ramped up outreach and interaction with the NWS and external weather forecasting communities to inform these current users of the advances that GOES-R data will provide and how it will assist weather forecasters. NOAA is using the Suomi NPP data to demonstrate the significant enhancements that these data are providing compared to the POES data currently being provided by NOAA-19 and NASA EOS satellites. This overlap of coverage allows users to characterize the utility of Suomi NPP data with the assurance of having the NOAA-19 and NASA EOS data on-hand to meet their immediate mission requirements.

We are cognizant that each day is a day closer to the eventual retirement of the current legacy satellites: the GOES-N Series, NOAA-19 and the NASA research satellites. This reality underscores the urgency that we must maintain the current development schedule to ensure that the first JPSS satellite is launched no later than the second quarter of FY 2017.

We rely on your continued support to ensure that these critical programs are fully funded in FY 2014 and beyond.

CONCLUSION

I will conclude by reaffirming that the NOAA-NASA partnership is strong and is supporting the good progress that the GOES-R Series and JPSS Programs are making. Stability of funding in FY 2014 and beyond is critical for meeting developmental milestones that will allow these programs to meet their launch dates. NOAA shares the concern of the Congress that gaps in coverage would significantly impact our ability to provide the weather coverage that the American public and commerce rely on. NOAA and the Department of Commerce thank the GAO for the very important contributions they are making to these programs. Mr. Powner and his team's recommendations offer us the opportunity for continuous improvement as we move forward to maintain the continuity of the operational environmental satellites that are so crucial to protecting American lives and property. Further, we express our appreciation to the members of the IRT and the OIG staff who have also provided significant and valuable review and recommendations. We accept their recommendations and will be responsive to them.

Finally, I wish to say that NOAA values the long-standing interest by the Committee and its staff regarding NOAA's satellite program. We understand the difficult fiscal environment that we find ourselves in and appreciate the Congressional support to ensure that these critical national programs are supported to the maximum extent possible. I am happy to answer any questions you may have.

In: Weather Satellites
Editor: Sherril G. Ryder
ISBN: 978-1-62948-687-1

Chapter 4

Statement of Marcus A. Watkins, Director, Joint Agency Satellite Division, National Aeronautics and Space Administration. Hearing on "Dysfunction in Management of Weather and Climate Satellites"*

Mr. Chairmen and Members of the Subcommittees, thank you for the opportunity to appear today to provide you information regarding the NASA role in, and commitment to, the National Oceanic and Atmospheric Administration (NOAA) Joint Polar Satellite System (JPSS) and Geostationary Operational Environmental Satellite-R Series (GOES-R) Programs. The JPSS and GOES-R Programs are critical to the Nation's weather forecasting system, environmental monitoring and research activities.

JPSS Organization Is Working Well

NASA and NOAA have been partners for more than 40 years in developing the Nation's polar and geosynchronous weather satellites. With the President's direction in 2010, NASA and NOAA included JPSS in this

* This is an edited, reformatted and augmented version of a statement, presented September 19, 2013 before the House Committee on Science, Space, and Technology, Subcommittees on Environment and Oversight.

partnership. The NASA program office for JPSS has been established and is fully staffed with a complement of 114 NASA civil servants and 310 support contractors. NOAA and NASA have established joint agency-level program management councils to oversee JPSS, and have integrated their decision-making processes to efficiently and effectively manage this cooperative activity. The NASA and NOAA teams have strengthened their working relationship over the last three years. The following are a few of our latest successes.

Suomi NPP Continues its Successful Operation

The Suomi National Polar-orbiting Partnership, or "Suomi-NPP," was developed to extend the record of key observations from the NASA Earth Observing System (EOS) series of satellites and to demonstrate space flight and ground data processing technologies for the next generation of operational polar-orbiting meteorological satellites. Suomi-NPP was successfully launched on October 28, 2011, activation and initial checkout are now complete, and NOAA has operational control of the satellite.

We are now reaching the end of a planned validation period, during which we are comparing the performance of the new sensors, both with data from on-orbit legacy instruments, and with high-quality ground-based and airborne calibration standards. As we characterize the performance of these new sensors, Suomi-NPP provides feedback to improve the operational instruments that will fly on JPSS.

While the satellite was not originally intended to be used as an operational asset, NOAA will use Suomi-NPP data in its operational weather forecasting models. NOAA meteorologists are already using data products from three instruments – the Advanced Technology Microwave Sounder (ATMS), the Cross-track Infrared Sounder (CrIS), and the Visible Infrared Imager Radiometer Suite (VIIRS) – in their weather forecasts. All of the data products have been publicly released, and our analyses are indicating that the instruments on Suomi-NPP are performing excellently.

JPSS Baseline Is Established

In addition to the successful Suomi NPP launch, the transition from the NPOESS program to the new JPSS program is now complete. The JPSS program successfully completed two critical milestones in July 2013, keeping the program on schedule and within budget. Both milestones were based on JPSS program content as outlined in the President's FY 2014 Budget request for a streamlined program designed to meet NOAA's weather mission. On July 17, 2013, NOAA approved the Key Decision Point –C for the JPSS-1 mission, establishing a baseline commitment confirming a second-quarter FY 2017 launch. On July 31, 2013, the Department of Commerce approved a combined Milestone 2/3 and Key Decision Point-I for the overall JPSS Program. These combined milestones established the full JPSS Program lifecycle cost of $11.3B, covering operations through FY 2025, the launch of JPSS-1 by second quarter FY 2017, and the launch of JPSS-2 by first quarter FY 2022. With these critical milestones now completed, both the JPSS-1 satellite mission and overall program move from the planning and formulation phase, to implementation and execution.

NASA, as NOAA's acquisition agent, manages all of the JPSS instrument, spacecraft, and ground system contracts. The first JPSS satellite, JPSS-1, will be a near clone of Suomi-NPP with upgrades to meet the JPSS Level 1 requirements. The instrument vendors continue to make progress in the manufacture of the flight units for JPSS-1, and the spacecraft is currently in development at Ball Aerospace. Additionally, the Delta II rocket has been selected as the launch vehicle for the JPSS-1 mission.

GOES-R Series Program Continues to Make Progress

The GOES-R Series Program of four geosynchronous satellites continues to make progress toward launching GOES-R. Last fall, the GOES-R Series Program successfully completed a Mission Critical Design Review. Since then, the GOES-R and S spacecraft have made good progress in component manufacturing, and GOES-R is proceeding with spacecraft integration. Four of the six GOES-R instruments have completed environmental testing. The GOES-R Extreme Ultraviolet and X-ray Irradiance Sensor instrument is in storage awaiting integration onto the spacecraft. The GOES-R Advanced Baseline Imager, Space Environment In-Situ Suite, and the Solar Ultraviolet Imager instruments have all completed environmental testing and will undergo pre-shipment reviews this fall in preparation for integration with the spacecraft next spring. The Geostationary Lightning Mapper instrument is completing its final assembly phase and will begin instrument-level environmental testing early this fall.

The next major milestone for the GOES-R Series Program is the Systems Integration Review, which is currently planned for the spring of 2014. This review will evaluate the readiness of the program to start system assembly, test and launch operations. Given the magnitude of the work planned for FY2014, any reduction in the requested GOES-R Series budget will have significant impact on the program schedule and life cycle cost.

Total Solar Irradiance

One example of the JPSS organization's success is the upcoming launch of Total Solar Irradiance Calibration Transfer Experiment (TCTE) later this year. The Total Solar Irradiance (TSI) climate record, which extends from 1978 to the present, is critical to understanding Earth's climate variability. The SOlar Radiation and Climate Experiment (SORCE) is currently providing data continuity for the record, but there could be a gap between the end of SORCE and the launch of NOAA's next TSI mission, the Total Solar Irradiance Sensor (TSIS). As such, the JPSS organization plans to launch the TSI Calibration Transfer Experiment (TCTE) late this year on a United States Air Force mission.

Conclusion

NASA and NOAA are committed to the JPSS and GOES-R programs, and ensuring the success of these programs is essential to both agencies and the Nation. The NASA and NOAA teams have established strong working relationships and are striving to ensure that weather and environmental requirements are met on the most efficient and predictable schedule without reducing system capabilities or further increasing risk.

With the launch of Suomi NPP in October 2011, the first fruits of the NASA-NOAA partnership for JPSS are being realized. With your continued support, NASA and NOAA expect this partnership to successfully develop and deliver the JPSS-1 mission on time for launch in FY 2017, thus ensuring the continuation of the Nation's capability to monitor the weather and environment.

Mr. Chairmen, I appreciate the continued support of these Subcommittees and the Congress, and I would be pleased to respond to any questions you or the other Members of the Subcommittees may have.

INDEX

A

B

C

D

E

F

R

S

T

U

V

W